Mustang Genesis

Mustang Genesis

The Creation of the Pony Car

ROBERT A. FRIA

Foreword by Lee Iacocca

McFarland & Company, Inc., Publishers
Jefferson, North Carolina, and London

LIBRARY OF CONGRESS CATALOGUING-IN-PUBLICATION DATA

Fria, Robert A., 1942–
Mustang genesis : the creation of the pony car /
Robert A. Fria ; foreword by Lee Iacocca.
p. cm.
Includes bibliographical references and index.

ISBN 978-0-7864-5840-0
illustrated case binding : 50# alkaline paper ♾

1. Mustang automobile — History. I. Title.
TL215.M8F74 2010
629.222'2 — dc22 2010025665

British Library cataloguing data are available

On the cover: Lee Iacocca and Don Frey with a specially licensed 1965 Mustang (Ford Motor Company); (background) original recreation of one of the first concept drawings done in 1962 for Ford by Gale Halderman of the Joe Oros Advanced Studio (courtesy of the artist)

Manufactured in the United States of America

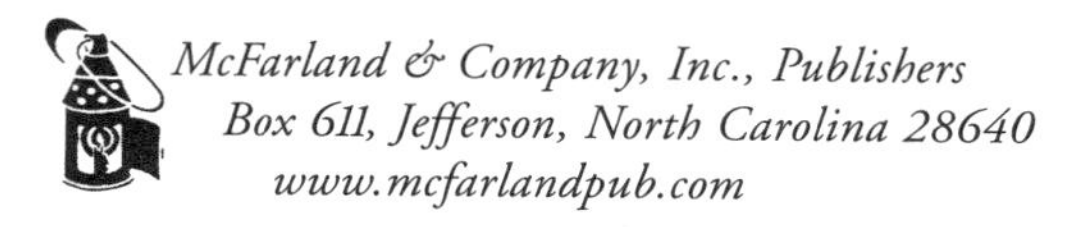

To Matthew
Our son who left this world
too young to become the next owner of the
first pre-production Mustang Hardtop

CONTENTS

Acknowledgments

Family comes first in my way of life, then everything else. My rock for 40 years has been my loving wife Joyce. She provided me with inspiration as she supportively and unselfishly helped me move forward with this book. She and our daughter Nicole have been my motivation. My thanks forever to both of you.

Writing a book is not easy, as I have found out. When that book is historical in nature, it is important and necessary to have the facts correct to preserve credibility. There have been many books published documenting the history of Ford Motor Company's Mustang. Information presented in some of them has been proven incorrect. I have tried here to present only documented information so that the origin of the car might be recorded in its purest chronological form, thus correcting previous myths and folklore.

It all began with my acquiring the first pre-production Mustang hardtop. In a quest to record the facts relating to this car I contacted many individuals who gave unselfishly of their time toward my research. Without help from these knowledgeable sources, my task would have been impossible. The research also led to my acquisition of the only known set of original blueprints for the 1962 two-seat Mustang.

Jim Smart, the walking archive of encyclopedic Mustang knowledge and caretaker to an immense resource of photographs, prodded me into this endeavor. "Don't leave our earth with this wealth of Mustang knowledge tucked away in your filing cabinet," he kept telling me. His desire to further the understanding of Mustang's beginnings through a published work was a driving force.

I first met Lee Iacocca in 2000 when he graciously posed for photographs with my car. Since then I have absorbed many facts about the process that created the car through subsequent personal interactions with him. I have been privileged to accompany him to several events and have been present at and assisted with some interviews and video recordings. Thank you, Mr. Iacocca, for allowing me into your private world and the opportunity to observe how a master thinks. And to personal assistant Norma Saken, thanks for opening the avenue to the stars.

Hal Sperlich took the time to verbally walk me through the entire Mustang project which he had managed. Once more, for history, this successful automotive giant gave of his time to help set the record straight. His shared insights open many avenues to a better understanding of the entire program.

I will be forever indebted to Gale Halderman for creating the concept sketch that is used on the cover of this book. His name and superior concept design are firmly immortalized in the Mustang. The interview with him is memorable as we covered the styling phase of the Mustang and new information was secured for history. And without his input, I would never have made many of the contacts with other retired Ford employees who provided previously untold information about Mustang development.

The eternal stylist of all things Ford, John Najjar, finally cleared up in first person the mysteries of how the first Mustang received its name. I was able from our discussion to gain a greater understanding of how Ford design came together to produce some of the finest styled cars of the century.

Bob Negstad opened his personal archives from his days as a Ford chassis engineer in order to assist me in documenting my car. Sadly, he has since passed away, and so will never see this expression of my gratitude to him in print. The Henry Ford's Bob Casey gave me access to inspect the first VIN-numbered Mustang convertible, which allowed me to prove many realities of early production.

Dennis Kolodziej, a Ford Fairlane Thunderbolt enthusiast, also provided invaluable early documentation establishing the identity of pilot vehicles.

Holly Clark is the daughter of Mustang I emblem designer Phil Clark. She is dedicated to ensuring that the world knows of her dad's achievements. She methodically went through her deceased father's personal files of documents and drawings and compiled them into a self-published book. It provides an in-depth look at a stylist's designs through the eyes of a proud daughter. I am indebted to her for making these documents available to me for my research.

Bob Gurr was the Disney supervising engineer who was responsible for the design of the Magic Skyway in the Ford Rotunda at the 1964 World's Fair. During our interview he provided an excellent description of the Skyway ride as well giving me an understanding of the introduction of the Mustang at the Fair.

Chuck Carter, Steve Grant, Frank Korn, Arnold Marks, Klaus Schaefer and Londoner Frank Middleweek are all Mustang aficionados and devoted hobbyists who contributed and shared their knowledge.

My thanks go to Jim Smart and Shelby Cobra guru Lynn Park for sharing photographs.

John Clinard, Western Regional Manager of Ford Public Affairs, has always been a friend. Thanks, John, for opening doors for me at Ford Motor Company, and remember my standing invitation to utilize the pre-production Mustang hardtop to represent Ford at shows and events.

Dick Clubb and Helen Hutchings provided needed technical support to the project. I also want to acknowledge all of the many others who, sometimes unknowingly, provided key pieces of history that made the jigsaw puzzle all come together into a recognizable whole.

Foreword

Lee Iacocca. *Lee Iacocca Archives*

There was a unique time in the twentieth century when our population was positioned toward a defined shift in attitude and lifestyle. The 1960s were those exceptional years, a time when a new way of thinking by the up-and-coming baby boomers spread into every aspect of the economy.

I was in a position to be able to influence the introduction of a new concept in automotive transportation. Perfect timing, in harmony with the baby boomers' new way of thinking, gave me the chance to assemble a special, talented team at Ford Motor Company with the specific goal of creating a new concept, four-seat sporty car for the young generation.

We hit the target dead center with the Mustang. From concept to production of those first cars, the history presented here by Bob Fria reflects the mood of the country and the interaction of the personalities who brought all of us the Mustang. We were able to create something good for America.

The Mustang has become an American icon, an association of which I have been very proud to be a part. America is a better place because of our concept named Mustang.

Preface

I started flying airplanes when I was fifteen years old. Somehow I knew flying would consume a major portion of my life, indeed even become my vocation.

It was a beautiful blue-sky day along the front range of the Rocky Mountains near Fort Collins, Colorado. A cadet in the Air Force ROTC Flight Training Program at Colorado State University, I was flying solo that day. As I was getting out of the Cessna after landing, a good friend came running up to greet me and announced that he was taking a demonstration ride in one of those brand-new Ford Mustang convertibles. I followed him out to the gleaming new top-down convertible, poppy-red with white bucket seats. It was spectacular. This pony was ready to prance. It was a ride that I have never forgotten.

Only a few weeks earlier, in mid–March, Ron McLean, the son of actor Jimmy Stewart and one of my best friends at the university, burst into the lunchroom with a big grin. "Bob, have you seen the new Mustang?" he asked as he waved a fresh copy of *Newsweek* magazine. There on page 80 was a sneak photo of the car and an article titled "Stray Mustang," announcing Ford would formally introduce the car in four weeks. I took in the sleekness of this new shape — long hood, flowing lines and a short rear deck that was enhanced overall by low body flair. The first words out of my mouth I remember clearly to this day: "This car will be a great success. They'll sell millions!"

And so they did. I didn't understand then that this car had been created by Ford Motor Company specifically for my generation. And though the marketing too was strategically developed to reach me and my peers, it would for me be a few years before I could afford to own one. Air Force commitments, then marriage and starting a family all came before my first Mustang — a well-used maroon 1966 convertible. The old-car hobby has kept me engrossed ever since.

In 1997, while I was skimming through an issue of *Hemmings Motor News*, one classified ad caught my eye. It was for a 1965 Mustang hardtop and mentioned that the car was one of the first ones built. I wasn't looking for another Mustang, but my curiosity got the best of me. Not only did I respond to the ad, I bought the car which turned out to be the *first pre-production Mustang hardtop* to be assigned a Vehicle Identification Number by Ford.

Initially I spent three years researching and developing the car's history so that I could restore it properly. There were many gaps in the information that was readily available. Gradually I gained insight into the creation of Mustang in general and my car in particular. The results of my ten years of research fill the pages of this book. By the time you've turned the last page, you too will have gained an understanding of what led to the creation of a little car Ford named after a horse.

Chapter 1

When Johnny Came Marching Home

To create a legend, first you have to set the stage. For the Ford Mustang, the stage was set at the end of World War II. GIs returning to the States from foreign lands had many new experiences to relate. Forgetting the hard times of war was difficult. Seeing a friend bleeding to death in one's arms was something one would never forget. The days of no food, the sleepless nights listening to bombardments, the thought of not knowing whether you'd be alive the next day, all hell-on-earth emotions set to the tune of anxiety. The wars in the Pacific and in Europe were long ones, with countless thousands of lives lost on all sides. Homes here in the U.S. were left without husbands, fathers, sons, brothers.

But there was good that came out of World War II. New economies were born, new industries were created, and America fell into unbridled postwar prosperity. During the war, between the years 1942 and '45, civilian automobile production in this country was eliminated to allow factory usage for military production of trucks, tanks, planes and boats. Large U.S. auto manufacturers found themselves in uncharted territory and were required to build needed mechanized war supplies. Smaller auto manufacturers went out of business due to constraints on purchasing materials needed for auto production.

The "Big Three" oligopoly, consisting of General Motors, Ford and Chrysler, had emerged after the great depression of the '30s. Ford Motor Company had become one of the largest manufacturers for military war supplies. Dearborn, Michigan, and the surrounding communities were heavily involved in the creation of boats, airplanes and all-purpose vehicles built by Ford. The four smokestacks at the famous Rouge Plant belched black smoke twenty-four hours a day. The clang of heavy metal processing could be heard afar. By 1943, the nearby Willow Run factory was rolling out thousands of B-24 Liberator bomber aircraft per year at the behest of Army Air Corps General Jimmy Doolittle.

Ford had made a major transition in its factories to wartime manufacturing despite an internal organization that was somewhat disorganized. Business was good: Ford had banked $5.25 billion in defense contracts by the end of the war.

But there was trouble on the horizon for the Dearborn company. The domineering architect Henry Ford had been pushing his only son Edsel hard. The future of Ford Motor Company would be left in his hands after Henry was gone. Edsel was named president of Ford at age 25 in 1919. By age 49, Edsel had developed a serious health problem. He developed what was thought to be a stomach ulcer, which left him largely incapacitated by 1943. He would succumb that year from undiagnosed stomach cancer. The man with just a high school education, and who would become known as one of the best stylists Ford Motor Company had ever known, was dead.

Henry was ailing and distraught when, at 80 years old, he was forced to return to running Ford Motor Company as president, full-time every day until the end of the war. Edsel had four children, three of them boys. Henry II was the oldest, Benson was next and William (Bill) was the youngest. Henry II and Benson had enlisted in the military and were serving the country when their father Edsel passed away. Bill was away at Hotchkiss School, a prep school in Connecticut. Edsel's only daughter, Josephine (known in the family as Dodie), had just married Walter Buhl Ford II (coincidentally, he had the same last name, but was not of the Henry Ford family). Edsel always referred jokingly to her side of the family as the "Ford-Fords."

Eldest son Henry II had always shown a keen interest in the day-to-day operations of the company, and was groomed by his father Edsel to become heir apparent of Ford Motor Company. Of the three brothers, he was deemed the strongest candidate to take over the company operations. Benson had dropped out of Princeton, and Bill was too

Henry Ford II. *Ford Motor Co.*

young at the time for consideration. On September 21, 1945, Henry Ford II was installed as the new president of Ford Motor Company.

In the same month that World War II ended with the official surrender of Japan, Ford's new king and his queen, Anne McDonnell, daughter of a prominent wealthy New York society family, arrived on the property. Henry II was granted formal early release papers from the United States Navy signed by Secretary of the Navy Frank Knox, as directed by President Franklin Roosevelt. It was deemed Henry II could better serve his country by running the behemoth Ford Motor Company than sailing boats for Uncle Sam.

Our story begins with Henry II firmly at the helm of this ship of fate. The following summer an aggressive twenty-two-year-old mechanical engineer, the son of Italian immigrant parents, who had just graduated from Princeton University with a master's degree, would be hired at Ford. His name was Lido A. Iacocca. Indeed, the stage had been set.

We need to know more about Henry II, who at his own insistence was always addressed as "Mr. Ford." He would become "Hank the Deuce" to some, but only behind his back. Understanding his makeup is essential since he figured so prominently in the origin of the Mustang. Born as Edsel's first child in 1917, he eventually rose to become Ford president in 1945 at age 28, then chief executive officer and chairman of the board before he died in 1987. One of his first very public appearances as the new Ford boss would be at the Indianapolis 500 in May of 1946. He drove the pace car, a 1946 Lincoln Continental V-12, conceived by his late father Edsel. His career persona ranged through born innocence, brilliance, organized chaos, and philanthropy to create an aggressive management style that led to his one-sided control of the fourth-largest industrial corporation in the world. Henry was complex and crazy like a fox.

A young Mr. Ford at work. *Ford Motor Co.*

At the end of the war, Ford Motor Company found itself in disorganization. Henry Ford, America's leading industrialist, had died at his home on a rainy, windy day, April 7, 1947. Edsel, Henry's only son and Henry II's father, was also gone. Henry II was trying to hold together a company that had been transformed from automobile manufacturing into war-machine production. The need for the mass production of bombers, Jeeps, boats and tanks was finished. Ford plants were ill equipped to proceed immediately back into automobile, truck and tractor production.

The financial fortunes had been spent. Cars rolling off the reconfigured assembly lines were uninspiring, using early '40s styling. Domestic sales were declining. Henry wore a concealed, holstered gun to work due to perceived labor problems, even though he was warmly received on the property. Unbelievably, Ford Motor Company was still using a decrepit manual accounting system with old-fashioned bookkeepers stacking invoices in piles up to four feet high. Workers were estimating accounts payable by measuring the height of the stacks with a yardstick and guessing at the values.

In the spring of 1946, employees reported "Henry II breathed the spirit of life into the place." But shy Henry realized he needed organizational help. He became aware that winter of a group of ten young Army Air Force officer veterans, highly skilled in the areas of production and management, who were collectively team-marketing their management talents developed while running the Air Force Office of Statistical Control. Henry investigated the group and decided they were exactly what he needed to help institute proper corporate management within the company.

They were hired en masse on a take-all-only package deal effective February 1, 1946. At first the group was known within the company as the "Quiz Kids." They focused for the first four months only on asking questions as they moved from department to department. They used the information gleaned to develop a new operating strategy. Their unique managerial style gained acceptance with the workers. The name associated with the group evolved into the "Whiz Kids," indicating the degree of acceptance. Two of them would eventually rise to the position of Ford president.

Robert S. McNamara. *Ford Motor Co.*

One of the Whiz Kids, Robert Strange McNamara, became instrumental in Ford's top management in the '50s and later became Secretary of Defense for President John F. Kennedy. In his early prior career,

McNamara was a professor at the Harvard Business School. Lee Iacocca has since described him as "one of the smartest men I've ever met, with a phenomenal IQ and a steel trap mind. He was a mental giant. And as a bean counter — a finance man — he was as good as they came in his analytical skills. He was a computer before they had computers."

The Whiz Kids' concept of going outside the company to hire new designers, stylists and engineers brought about new designs that would once again breathe life into Ford cars. Those cars would become known as "the cars that saved Ford," beginning with the new 1949 Ford. Of note, the distinctive grille design on the '49 was done by a young stylist named Joe Oros, who worked for the design consulting firm Walker Industrial Design Co. This company acted as a consultant to Ford and is credited with the final design of the '49 Ford. Oros would be credited later personally with the design of the 1965 Mustang. The Whiz Kids were given great credit for leading Ford back into prosperity.

In mid–1945, Henry's middle brother Benson arrived at Ford after leaving the Army.

1949 Ford with Henry II (driving), Benson and Bill Ford. *Ford Motor Co.*

Benson's tenure at Ford was primarily involved with company administrative functions. He became vice-president and general manager of the Lincoln-Mercury Division in 1948.

Henry's younger brother Bill added to the Ford family in 1947 by marrying twenty-two-year-old Martha Firestone, granddaughter of the tire magnate Harvey Firestone. He would graduate from Yale two years later with a degree in economics and go right to work in Ford management.

Henry was looking for ways to boost sales. He visited war-torn Europe in 1948, the same year his son Edsel II was born, with an idea of how to breathe new life into his Ford Motor Company. Henry made an offer to buy 51 percent of Volkswagen, an offer that was quickly turned down and given little fanfare. It didn't seem to matter because the newly created Ford Division '49 Ford passenger car had just hit the streets.

It was wildly successful: sales were reported at 100,000 car orders on opening day, with sales helped by pent-up new car demand following the war. The successful sleek design of the Ford was new to its public, and that's what sold the car. The '49 Ford was later called "a piece of junk" by many consumers who said the body panels didn't fit, it leaked water, and it had other defects.

Modelers who worked on the design of the '49 had divided themselves into two reli-

1948 MG TC Roadster. *The author*

Typical 1950 Mercury customized coupe. *The author*

gious sects, the Catholics and the Protestants, according to stylist Gene Bordinat. Between unfriendly gestures and conversations shared resentfully between the two, they managed to get the car done. Despite these problems, with its new styling the '49 would be the flagship to carry the company forward into the fifties and save the company from financial ruin.

In 1949, Ford, Mercury and Lincoln lines would produce one million cars. By contrast, the not-so-lucky Tucker Corporation ceased making the Tucker car that same year after only 51 cars had been built. Ford was then in third place, only slightly behind GM and Chrysler. That year they made $177 million in profits. Henry's first new Ford cars had launched his success into the future and saved the company.

After the war, returning soldiers brought with them an affinity for the new-found European two-seat sports cars. Long, low and sleek, these affluent-owned roadsters like the British Jaguar and MG TC, the Italian Alfa Romeo, and the French Bugatti and Delahaye were a new concept to which our young GIs had not been exposed in the U.S. There are two things which catch the fancy of almost every young red-blooded American male: appropriately endowed females and a flashy new car. The die had been cast for U.S. auto manufacturers to supply the latter of these two head-turners.

These young guys wanted action. They found their auto passions headed toward built-for-speed hot rods, Ford customized V-8s powering their dry-lakes racers, and James Dean–style tub-like modified Mercurys of the late '40s and early '50s. These guys wanted speed, they wanted horsepower and they wanted open-air roadsters. Only problem was, none of the big auto manufacturers were building anything close to that description. If that's what turned you on in an automobile, then you would have to build one for yourself— and they did.

By the early '50s, streets across the U.S. were replete with custom-built hot rods, roadsters, and fire-breathing open-air passion pits on wheels. Drive-in theaters, benefiting from the movement, had doubled in number in this country from 1,100 in 1949 to 2,200 just one year later. Economic times were good, very good. New homes were being built by the thousands. There was a new network of superhighways to span the country and Ford had a great lineup of family cars in its Lincoln, Mercury and newly established Ford divisions.

But there were no two-seat sports cars in that lineup.

Chapter 2

In the Beginning There Was ...

"Just remember, it's my name on the side of the building."
Henry Ford II

The good, the bad and the ugly. No new cars were built in this country during World War II and it wasn't until 1949 that postwar automotive styling began to hit the marketplace. The *New York Times* reported the president of General Motors Corporation forecast an annual record production of 6,000,000 motor vehicles for the industry in 1949. It became a banner year for vehicles produced. However, the first years of the '50s brought many leftover ideas from those 1940-generation cars. Fat fenders, soggy suspensions and poor performance proliferated. Things had to change.

Notable designers were beginning to leave their strokes of brilliance on various marques across the industry. The genius of Harley Earl gave us the beginnings of the tailfin era at General Motors. His protégé Virgil Exner, now at Chrysler, was dreaming of aerodynamics and "Flite-Sweep" styling and had started design of a new car to be called the Chrysler 300. Unique styling from Ford's new Advanced Studio under the direction of A.G. "Gil" Spear utilized the talents of independent design consultant George Walker to develop the styling for the new 1949 Ford, and it was this innovative styling that continued into the early '50s at Ford.

All this combined styling gave us new concepts in sedans, station wagons and two-door hardtop coupes and convertibles. Ford introduced the Victoria pillarless coupe in 1951, a first for Ford. In 1952, Lincoln and Mercury both introduced true hardtops to their lines. A new way of styling was indeed introduced into the new decade. Since the war, the pipelines had been filled with new automobiles. A drop in demand for new cars in the U.S. came in 1951 and '52. Market saturation had peaked on top of a poor economic climate due largely to the buildup of strategic materials for the Korean War. The automobile market for new cars needed a kick start. Where were the sporty cars those World War II guys were clamoring for?

Sporty cars were made in Europe, so it was only logical for Ford to look there for new ideas to pump up U.S. sales. In 1950, 21,287 cars were imported into the United States, and by the end of the decade that number would climb to 608,070. By the early '50s, the Europeans were specifically designing sports cars for the Americans named Jaguar, Austin Healey, Alfa Romeo, and Mercedes. Even the Volkswagen Beetle went from 600 sales in 1954 to 30,000 in 1955 with the formation of the Volkswagen of America dealer

1951 Ford Victoria Hardtop. *Courtesy Kenneth R. Bounds*

network. A formidable new market had evolved and had to be recognized by the domestic auto industry.

Americans were fascinated with bucket seats, floor-mounted manual shifts and two-seat bodies. With those seeds planted, Henry II decided to own a sports car of his own to see what all the clamor was about. He always particularly liked cars from the Italian school of design. He was approached in 1949 by the principal owner of a small Italian car manufacturer named Cisitalia, of Turin. Their car, the Cisitalia 202, was a sleek two-passenger sports car just being imported into the United States.

It had caught the Chrysler stylists' eyes with its new emphasis on aerodynamics.

The Cisitalia Company was in need of an immediate money infusion. At first shunned by Chrysler for financial support, company executives approached Henry on incorporating the Cisitalia import line and proposed the use of Ford engines, transmissions and suspensions in the imported 202 model. Henry liked driving the car and thought it was what a well-designed sports car should look like, especially since popular Italian-American singer Perry Como was also interested in the car.

Henry drove one of the cars to the Ford Design Department lot one day and asked his designers why they couldn't design a similar car. Embarrassed, they couldn't give him a satisfactory answer. Upon Henry's departure, a young stylist named John Najjar (pro-

1952 Jaguar XK-120. *The author*

Cisitalia 202 Coupe.

nounced "nā-jār") was given the task of writing an answer to the Deuce. Henry never liked the answer he got in the report, and neither the car nor its styling was integrated into the Ford lineup.

Henry brought Lewis Crusoe, age 50, out of an early retirement from General Motors and a short stint at Bendix. Assigned as general manager of the new Ford Division, he was considered by many a "certified automotive genius" and believed in conservatism. His face was framed by thick rimless glasses. Crusoe formed an alliance with George W. Walker, a tough-talking independent designer who had introduced a fresh new look to some of the then-stylish cars for Nash Motor Company. Walker's Industrial Design Company was brought in as a consultant to the Ford Division and Walker himself began a professional relationship with Crusoe.

Much of the design responsibility for the '49 Ford came from Walker and his stylists brought to the Ford property for that purpose. Henry held George Walker in high esteem after the success of the '49 Ford. Walker spent much of his company's time and resources as a consultant to Ford, but returned to his own business of industrial design in 1949. He

George Walker, in the Ford Design Studio, 1955. *Ford Motor Co.*

continued to do consulting work for the Ford styling studio. In 1955 he was lured back to Ford Motor Company and was hired permanently as a full-time employee.

Colorful is a word inadequate to describe Walker. He appeared on the cover of *Time* magazine and was referred to as the "Cellini of Chrome" for his glitzy, adventuresome, ahead-of-their-time designs, which suggested his sometimes in-your-face style. Described as a blunt, crude, tough-talking man, he was absolutely merciless to those who didn't see things his way. Walker was the styling department force to be reckoned with on any new design.

In October of 1951, Crusoe and Walker went to the Paris Auto Show in search of new ideas. As the two walked the aisles of the Grand Palais in Paris, Crusoe couldn't help noticing the low-slung, streamlined sports car models on display. Manufactured by prominent companies such as Jaguar and Bugatti, the cars stood apart from all other models on display. Wondering aloud, Crusoe said, "Why can't we have something like these cars?" Walker quietly and immediately placed a transatlantic call to the Ford Division Styling Department in Dearborn.

Walker and Crusoe initiated a project to create such a styling design, and upon their return from Europe, they developed their first thoughts toward creating a new Ford two-seat sports car to become known as the "Thunderbird." It was in 1952 the first paper shaped two-seat sports car design was taking place in the styling department. That's about as far as it went as all eyes became necessarily more focused on preliminary designs for the newly styled Ford 1955 model year cars. The paper designs of the two-seater were relegated to the designer's drawer.

General Motors was the first major manufacturer to realize there was indeed a market for a new two-seat auto design. New direction was given designer Harley Earl heading the GM design department and their design idea in May of 1952 would produce the first entirely new two-seat mass-produced roadster-styled sports car in America. A full-scale plaster model was built with the name "Corvair." This roadster was shown to the public in January of 1953 at the GM Motorama held at the New York City Waldorf Astoria hotel. They built it in a new revolutionary material called "glass-fibre." Upon its introduction at the Motorama show, the formal name had been changed to Corvette. In little more than six months, the first production unit rolled off the GM assembly line in July 1953, with an anemic "Blue Flame" 6-cylinder engine.

Barely three hundred Corvettes were built that year by GM, but they had sensed a pulse for this style of car. Total production numbered fewer than thirty-seven hundred cars sold through 1954, before a V-8 engine was introduced into the restyled Corvette in 1955. Now car buyers finally had a roadster from GM with two seats and a V-8 engine to light the smoldering fire under that pent-up demand Johnny brought home after the war.

Where was FoMoCo? In 1952, the Ford styling department was under the supervision of Franklin Q. Hershey, a new employee. Raised in Beverly Hills, he had majored in forestry at Occidental College in Los Angeles. As an auto designer, he devised skillful styling that was incorporated into Duesenberg and Packard roadsters, and he later became

1953 Chevrolet Corvette. *Courtesy ProteamCorvette.com*

a student of Harley Earl at GM. There he was credited with having originated tailfins for the Cadillac line beginning in 1948, an idea he said came from a famous World War II fighter, the P-38 Lightning twin-tail boom aircraft.

Clandestinely, an old friend of Hershey's working in the GM styling studios turned over photos of the newly designed concept Corvette to him. Reaction to these photos caused the plans for the concept car project Crusoe had begun in 1951 to be pulled out of the drawer and dusted off. Hershey's department then proceeded with development. Hershey made secret drawings of what he conceived as a Ford answer to this new two-seat phenomenon, unbeknownst to higher Ford management. Not waiting for upper-level approval of a new project, Hershey's department, with the help of assistant Eugene Bordinat, continued with the development of Ford's answer to the new glass-fiber car.

When GM introduced the Corvette, Henry II decided it was time to investigate this phenomenon and jumped in the arena on February 9, 1953. He issued a directive to proceed full speed ahead with development of a new Ford concept car. He had no idea there was a design already lurking in the bowels of his own design department.

The year 1953 would be a busy one for Mr. Ford. A major celebration commemorating the 50th anniversary of the Ford Motor Company was planned. For the event, he had a medallion struck featuring the joined profiles of Henry I, Edsel and himself denoting "old age, middle age and youth, over fifty years." This medallion affirmed his place in American respectability. The realized result of this and of the 50th anniversary celebration was that Henry II had now officially inhabited the Ford legend.

President Dwight Eisenhower appointed Henry II, as a major business leader, an alternate delegate to the United Nations. "I went to the UN to learn about the world," said Henry. "I was just a kid. I didn't know anything about the world. Ike never paid any attention to me, but he did ask me to go to the UN." This was his first real experience with big-time politics.

Plans progressed toward a new styling concept according to Henry's directive. Hershey's department used the 1953 Corvette concept, an MG TD, a Jaguar XK-120, an Aston Martin, a Ferrari and a Nash-Healey as study examples for the new design to become known as Thunderbird. There were few secrets in the industry, with GM, Ford and Chrysler spying daily on each other's styling. When the Ford Design Department moved to its new building on Oakwood Boulevard in Dearborn in May 1953, managers never noticed a two-story house that sat on a knoll nearby. Anyone on the second floor of that house had a clear view of the Ford Design Department's enclosed courtyard where new concepts, including the Thunderbird, were displayed. One of the competitors rented the house and was using it to spy on Ford's designs. As soon as Ford figured out there was a "leak" and where it was coming from, they bought the old house and immediately had it torn down.

There was never mention of the Ford Competitive Analysis Department that existed in the early '50s. Some of this department's projects, managed under future Fairlane Committee member Chase Morsey, included sending Ford spies with cameras strapped to their necks to climb fences in Arizona at the GM Proving Grounds to see what they could find.

First public viewing of the new two-seat Thunderbird was made at the Detroit Auto Show on February 20, 1954. The final fiberglass styled model was displayed. Initial reaction to the car was very positive and plans were made to set up retail production. Thunderbird went on sale to the public October 22, 1954, as a 1955 model with a retail price of $2,695, $5 less than a new 1955 Corvette. The new Corvette had sold poorly prior to 1955, and was almost killed by GM management for lack of sales. It was the new Ford Thunderbird with its V-8 and better styling that forced GM into their new restyled Corvette, which saved it from extinction that year.

The new Thunderbird roadster with its removable tops and roll-up windows received its name through an internal Ford contest with employees submitting suggested names. Stylist Alden Giberson submitted the winning name "Thunderbird" and was awarded the grand prize of $250 toward the purchase of a new men's suit. A fiscal conservative, he purchased a new $95 suit with an extra pair of pants at Saks Fifth Avenue. The Ford Division needed some prodding but eventually did pay for the suit, weaseling out of their original $250 award. Giberson never complained. Franklin Q. Hershey ultimately received credit for the design of the 1955 Thunderbird.

On Thunderbird's introductory day, October 22, 1954, Ford dealers took orders for an astounding 4,000 cars. Movie actress Jane Wyman agreed to host a party with a new Thunderbird in her living room if she could buy the first new car. For the 1955 model year, Ford sold 16,055 Thunderbirds while GM sold only 674 Corvettes. By the end of the 1956 model year, Ford had sold 15,631 of the '56 two-seat roadsters. The redesigned 1957 T-Bird sold 21,380 units after an extended year run. The two-seat Thunderbird era

1955 Ford Thunderbird. *Ford Motor Co.*

ended with total combined sales for all three years of 53,166, compared to the 14,446 Corvettes sold over the five-year run from 1953 through 1957.

Karl Ludvigsen, in an article published in *Automobile Quarterly*, described the Thunderbird this way:

> It is an outstandingly handsome car entirely in the American idiom, a collector's car that's also practical and enjoyable for daily transportation and long trips, a stylish and selfish sporting machine with excellent performance and an intriguing pedigree. It turned out that way because it was brought to life by men who knew what good cars were at a time when the company they worked for needed something good—and fast.

That comment was written in the fall of 1970, and the assessment remains true today.

Before the first public introduction of the 1955 Thunderbird in the Ford Rotunda building in Dearborn, Ford Division General Manager Lewis Crusoe informed product planner Tom Case that although the new T-Bird was beautiful, it needed two more seats. Henry II let it be known that he loved his new Thunderbird but he couldn't get his golf clubs in the small trunk and it was a bit embarrassing in front of his golfing friends. Immediately plans were made to start on a new four-seat design, a concept that was pushed by new Ford Division General Manager Robert McNamara with a team that would eventually include a new Dearborn transferee from the east coast named Lee Iacocca.

The idea was to create a new four-seat sporty car with more market appeal than the two-seater. Even as the final plans were approved for the '56 T-Bird and styling mockups were being shown for the '57 T-Bird in March of 1955, there was a full-size clay mockup for the "195H program" that would produce the four-seat 1958 Thunderbird. The final '58 product would be constrained by the size and weight of the big and heavy performance engines available at the time. Engine designers came up with a new 221-cubic-inch V-8 in 1961 that was designed with so-called thin wall technology that made it much smaller and lighter. "This size engine then made it possible to design a four-seat sporty car like the Mustang. The four-seat Thunderbird for 1958 is really what the Mustang would have been if that engine technology was available," said Tom Case, Thunderbird project manager for the Ford Division.

Gone by the end of the decade were auto manufacturers Packard, Hudson, Nash and Kaiser, companies unable to come up with what America wanted. Additionally, Chrysler Corporation's DeSoto was on the very quick trip to extinction.

The two-seat sports car era at Ford was done, but it would not go unnoticed. Ford's corporate sales office was inundated with letters and inquiries calling for the rebirth of a true two-seat sports car. The first and only true two-seat sports car still being produced in America was the Chevrolet Corvette, and devoted Ford loyalists were feeling left out.

Associated with the design and styling of that original two-seat Thunderbird in 1954 was a group of men who worked in styling departments at Ford. Remember the names Gene Bordinat, Joe Oros, Dave Ash and John Najjar. Ten years later, in a corroborative effort, these men would design and build one of the greatest cars Ford would ever produce.

1958 four-seat Ford Thunderbird. *Ford Motor Co.*

Chapter 3

A Decade of Decadence

In 1952, Dwight Eisenhower, former supreme commander of the Allied Forces in the European Theater of Operations in World War II, was elected president of the United States. The cost of a new Ford car was $1,800 and it ran on $.20-per-gallon gasoline. At the end of the Korean War in 1953, it was time for consumers to resume robust spending as national prosperity returned.

This would be the revolutionary decade of car design for the "Big Three" auto manufacturers. Institutionalized in retail design were power steering and brakes, power seats and windows, glass and retractable tops, one-piece windshields, and lots, yes, *lots* of chrome. Henry II would accomplish many positive things at Ford Motor Company by the end of the '50s.

A young PhD engineer named Don Frey was hired at Ford in 1950 and would later become instrumental in the creation of the Mustang car. In 1951, a University of Michigan graduate engineer would also join Ford. Hal Sperlich was later credited with the concept of using existing Falcon parts to build the new Mustang car. Without his concept, the Mustang project would probably have collapsed.

Beginning with the 1952 model year, the entire lineup of cars at Ford was restyled. By 1953 Ford Motor Company had overhead valve V-8s and four car lines with fourteen models, compared to two lines and seven models in 1949. By the end of the decade the new two-seat Thunderbird had outsold the GM Corvette three to one. A formidable attack against GM was now in full swing with Ford Motor Company gaining market share over GM and Chrysler. Henry II had resurrected Ford from the ashes of World War II and was the leader of one of the most powerful industrial companies in America.

The company was reaping good financial rewards from sales in the mid–'50s when Henry made a decision to offer Ford Motor Company stock to the public. Two classes of stock had been created in 1936 by Henry I and son Edsel, largely to avoid taxes when either of them died. The larger block was Type A, a nonvoting class. A smaller class block of Type B stock with voting privileges would control the company and was kept within the Ford family. They also created the Ford Foundation for philanthropic purposes, and willed all the Class A nonvoting stock to the foundation. Over the subsequent years, the stock given the Foundation grew to immense value based on the success of Ford Motor Company. It was second only to the Rockefeller Foundation, one of the country's major philanthropic foundations. Management at the Ford Foundation believed that owning only 88 percent of

Ford Motor Company was not enough, based on the annual dividends being paid to them.

They formed a committee to seek authority to sell shares of their common stock on the open market. This seemed fine with Henry II since he personally seemed always cash starved and he too was unhappy with the amount of dividends he was receiving from his stock. Family discontent ensued with the major sticking point being the ratio of common stock to Class B stock, which would determine what the Ford family percentages would be. Details were worked out and the biggest stock offering ever in America occurred in 1956. This single event, which Henry II engineered, was timed perfectly, and money began overflowing the floodgates. Ford Motor Company had money to spend, and spend they did. This was a turning point for the fortunes of Ford, both the company and the family. After twenty-five years, Ford had begun to move ahead of Chevrolet in annual sales.

A significant event occurred in the summer of '56 at Ford Corporate Headquarters in Dearborn that would ultimately change the course of the Ford Motor Company. Lee Iacocca left Dearborn in the late '40s as an engineer to pursue a career with Ford in sales rather than engineering. He headed to the east coast, where he eventually worked his way up to assistant sales manager in the Ford Philadelphia District sales office. He created a blistering sales program in 1956 which used the slogan "56 for '56"; it was an instant success, leading to the sale of an additional 75,000 Ford cars at a time when car sales were slow. The slogan meant a customer could buy a '56 Ford for $56 a month. His sales successes were noticed by the head shed in Dearborn and he was promoted that spring to Ford district sales manager of Washington, D.C.

The year 1956 would be career- and life-changing for thirty-two-year-old Iacocca. He would be married to Mary McCleary in September. They had met eight years earlier when she was a receptionist at the Ford assembly plant in Chester, Pennsylvania. The two had set a wedding date for that September and had just bought their first house. One week before the wedding, he was notified he was being transferred back to Dearborn. McNamara had recognized the results of his extremely successful sales campaign for the 1956 Ford cars.

He would have an office at the new Ford World Headquarters Building that was completed and opened for business on September 26, 1956. The multi-storied executive building on Michigan Avenue would tower above all others in Dearborn and was the perfect place for King Henry to command his troops — from the top floor. His area was filled with executive-level offices and a complete executive dining facility with a senior chef. The building and surrounding complex would later be renamed the Henry Ford II World Center.

Ford Motor Company began a campaign with the 1956 model year cars to offer optional safety-related devices for their cars. Some of these included padded dashes and sun visors, "deep dish" design steering wheels and seat belts. Only 2 percent of the buyers that model year opted for these safety devices. Safety education might have been emphasized better as one customer complained, "[The seat belts] were bulky and uncomfortable to sit on." Seat belts would not become federally mandated until the 1964 model year.

Creative designer Joe Oros, who had come back to Ford in 1955 along with George Walker, is credited with the beautifully designed '57 Ford car styling as well as the '58 Thunderbird styling. A young designer named Gale Halderman, also hired in 1955, was immediately put to work on the design of the 1957 Ford cars along with Franklin Hershey and Joe Oros. Halderman later would be one of the significant designers of the 1965 Mustang while again working with Oros. The '57 design was outstanding, but the economy was bad in the '57 and '58 model years, and the cars were not selling as had been hoped.

One major engineering development of those years stood out. It utilized the '58 Lincoln car chassis. A body-without-a-frame concept would become a new state of the art manufacturing process for Ford known as "unit-body construction." This construction method was also incorporated into the '58 T-Bird, with both these cars using assembly line commonality and both produced on the Wixom, Michigan, assembly line. John Najjar was head of the Lincoln styling studio from 1955 through '57 and was responsible for the design of the "bigger is better" era for the '57 and '58 Lincolns. These creations were not popular selling designs. Najjar would later be credited with the design of the Mustang Experimental Sports Car and assisted with styling of the production Mustang in 1965.

Bob Negstad was a promising young engineer from Oregon State University and was hired in 1958 right out of school. He drove his '57 T-Bird to Dearborn to start work with Ford Motor Company. He began his career specializing in the development of suspension designs. Negstad would later be instrumental in the suspension design of the original Mustang Experimental Sports Car and the '65 Mustang as well as the GT40.

The four-seat 1958 T-Bird envisioned by McNamara sold well through the 1960 model year, but it appealed to the more affluent country-club crowd. Even though it outsold the smaller two-seat version by three-to-one its first year, it was publicized and perceived as a luxury car and did not fulfill the emerging market demand of youthful buyers.

Overconfidence abounded within the halls of company headquarters in the mid-fifties, a very important factor to recognize. It would lead to overzealous production of a new model line to be known as the Edsel with its own special products division. There would be many new Ford, Mercury and Lincoln exotic models for the late '50s. Henry backed the Edsel project wholeheartedly. Many of his soldiers warned Ford was moving into dangerous territory by offering a new car that had no obvious styling or option benefits over similar GM lines. Henry overrode the objections, took the plunge, and approved the Edsel project on April 15, 1955.

The Edsel never had the blessing of McNamara, and he was not interested in providing any assistance to ensure the success of the car line. With the Edsel introduction date approaching, press coverage was high. On August 28, 1957, a dinner dance was scheduled for the press for the car's introduction. As authors Collier and Horowitz report in *The Fords*, at the dance that evening, Fairfax Cone, head of the advertising agency handling the new Edsel account, asked McNamara about his thoughts on the new car. In his staunchly conservative style, he simply replied, "I've got plans for phasing it out."

The year 1958 approached with car sales down due to a slow economy, and Ford

1958 Edsel with William, Benson and Henry Ford at the Edsel introduction. *Ford Motor Co.*

stock was priced low. McNamara insisted a $200 per car retail price increase on all Ford Motor Company cars be passed on to the buyer. Interestingly, he decreed the increase would first be applied to the Edsel, some three months before becoming effective on the other car lines.

Congress passed a new law known as the Monroney Act, which required a manufacturer's suggested retail price sticker be placed on a window of every newly manufactured car. The law became effective in the summer of 1957 and the Edsel was the first new 1958 model car required to have a Monroney sticker. This would be the first public exposure to how much retail prices were artificially inflated for profit motive by the retailer. Prospective Edsel buyers received their first dose of "sticker shock" and the cars were from the start perceived as being overpriced.

Critics overlooked the Edsel's revolutionary interior styling and concentrated on the grille design, created by designer Jim Sipple (later involved with Mustang design), which some compared to a horse collar, a man sucking a lemon and a toilet seat. One critic even said it should be called the "Ethel" and not "Edsel." An hour-long TV show on October 13, 1957, called *The Edsel Show*, which pre-empted *The Ed Sullivan Show*, with Bing

1958 Edsel.

Crosby and Frank Sinatra, was not enough to jump-start Edsel sales. By December, Henry II was pleading with dealers to hold onto the Edsel line. At the same time, McNamara was calling for the early termination of the Edsel to stop the flow of red ink. The car that Benson Ford had so adamantly campaigned against for using his father's name was in fatal trouble. On November 19, 1959, after only 84,000 cars had been sold, the Edsel was discontinued. With a loss of $350 million and a drop in Ford stock value of $20 per share, the defeat was costly.

Two persons who managed the Edsel project bore the responsibility for the disastrous Edsel's early demise. Lewis Crusoe, who had been head of the Ford Division, suffered a massive heart attack and retired. Jack Reith, who headed the Mercury Division, left Ford after the Edsel embarrassment and died shortly thereafter. The financial loss would be permanently etched into the minds of those involved with the project, especially Henry II. This would repress early attempts to gain his interest in building another yet-to-be-conceived model, which became a little car called the Mustang.

Ford acquired the Lincoln name from Henry Leland in 1922 and up until the mid–'50s, Ford had never made a profit on the car line. Sales of Lincolns in 1956 were strong, thanks to their sleek new styling makeover, and the line came close to producing its first division profit for Ford.

About this time, Henry's brother Benson, then head of the Lincoln-Mercury Division, was replaced. He was taken out of his position by Lewis Crusoe, a move unopposed and orchestrated by Henry. In 1957 Benson suffered a heart attack and was relegated to a job with the policy committee. A highlight for Benson came at the 1964 Indianapolis 500, where he drove the newly introduced 1965 Mustang convertible pace car for the race. He died from a heart attack in 1978 while cruising on the Cheboygan River on his 85-foot yacht.

Robert McNamara, head of the Ford Division, chose a path somewhat uncharacteristic for him by agreeing to go ahead with the new, non-austere concept car, the four-seat Thunderbird. Although it can be speculated and supported by sales figures that the Thunderbird needed to grow into a four-seat version to improve sales, there was more of a political motivation behind giving the OK to proceed with the car.

Ford Division chief McNamara wanted to provide stiff competition for the new Mercury Division models and the newly created Edsel Division cars. There was no way the two-seat T-Bird with limited production was going to provide that in-house competition to make the Ford Division shine over the other two divisions. In 1956, with no consideration of input from any others in the division, McNamara gave the order to stop production of the two-seat Thunderbird model. Tom Case, product planner for the two-seat T-Bird, had argued with McNamara about killing the car. After all, it had captured the imaginations of car buffs everywhere and was a great showroom attraction for the entire Ford line.

McNamara insisted that he would let nothing stand in the way of the new concept four-seat Thunderbird to be introduced in the fall of 1957 as a '58 model. He needed public focus on the new four-seater with no distractions from the "little Bird." His final declaration—"The car is dead and I don't want to hear another thing about it"—was the end of the magnificent two-seat Thunderbird, a car that went on to become known in survey after collector car survey as the most popular collector car of the twentieth century.

In late 1956 McNamara agreed to produce 100 1957 Ford production cars with superchargers so the company would meet qualifications to compete in the NASCAR racing circuit at Daytona Beach, Florida. This was detailed in an internal memo titled "1957 Ford Motor Company Supercharger Program" which he presented to the Ford executive committee. It included 65 sedans, 20 convertibles and 15 Thunderbirds, as a minimum. The plan was to continue the supercharger option into the new '58 Thunderbird and Ford car lineup, with as many as 5000 cars envisioned. An estimated total of 1500 cars from the Ford Division would actually be built, including 211 Thunderbirds.

Even before the '57 production car lines were closed, he decreed in June of '57 that Ford would no longer participate in high performance or racing events, due primarily to the Automobile Manufacturer Association (AMA) ban on factory-sponsored racing. He eliminated the supercharger option from all Ford cars. After all, Robert McNamara was an accountant, not a car buff. He was interested only in numbers and the profit and loss statement. He knew the two-seat T-Bird was a loser measured by number of sales. He

1957 Ford Sunliner. *The author*

decreed Ford would no longer build loss-leader cars, regardless of their sales potential. Instead everything Ford built from then on was going to be designed primarily to make money. If the car was also interesting, so much the better.

That was the case for the '57–'59 Ford Skyliner retractable hardtop cars. In '57 McNamara axed further production beyond the 1959 model year, even though it was a very interesting concept. Now, along with the demise of the little Thunderbird, he had one more obstacle standing in the way of his plan to make his Ford Division shine, and that was named the Edsel.

There were a lot of internal politics going on within the Ford Division in the late '50s. The "Big Plan" had been propagated to increase Mercury sales and introduce the Edsel line, mainly to give Ford Motor Company more of a car-line-for-car-line comparison to GM. There was a Ford Division, a Mercury Division, an Edsel Division, a Lincoln Division and a Continental Division. The next thing anybody knew, the Continental and Edsel Divisions were gone and the Mercury/Edsel/Lincoln Division evolved, if only for a very short time. Henry II brought in an outsider from Packard named John Nance to head that division. Nance couldn't even drive a car and was named one of the worst automotive executives ever. His tenure with Ford was very short-lived.

1956 Continental on the Arizona Ford test track. *Ford Motor Co.*

William Clay Ford Sr., more commonly known as Bill, and the youngest of Edsel's three sons, graduated from Yale in 1949 with a degree in economics and went to work at Ford for oldest brother Henry. His first top-level management job was as the director of the newly formed Continental Division with a duty to produce the first new executive-styled 1956 Continental cars.

The beautifully European-styled Continental two-door luxury car would be the most expensive new car in America, priced at $10,000. This was a handmade car designed for the elite buyer. First cars hit the retail market in 1955 as 1956 models. Production ran through the 1957 model and then ceased for lack of sufficient sales. A combined total of 1749 cars were built. Ford lost money on every one of the new Continentals due to high production costs.

Henry was upset with the losses. In a carefully calculated way that only he was capable of achieving, he arranged at arm's length to have the Continental Division disbanded and folded into the Lincoln car line. When Bill Ford vacated his position with the Continental Division, he pursued his interests with the Detroit Lions football team franchise, which he purchased in 1964, and still owns to this date.

Sweeping changes had occurred at Ford Motor Company during the decade. Even though the cars of the '50s from Detroit were large and heavy, there were lasting engineering innovations developed on them that we still use today, such as automatic transmissions, power steering, power brakes and new unibody chassis structures. Overall handling and driving were much improved. Ford had built its 25 millionth V-8 engine, which was installed in an Edsel. The mid-to-late '50s to the early '60s was a period in which a great deal was accomplished in the history of the American automobile.

And, in 1959, the Iacocca family welcomed their first daughter Kathryn into the world.

Chapter 4

THE PLAYERS

"Management is nothing more than motivating other people."
Lee Iacocca

As the transition was made into the decade of the '60s, in place at Ford Motor Company were the men who would magically become instrumental in the conception, design and development of those historically important cars that evolved from that performance era. The interactions and synergies between these men are what defined some of Ford Motor Company's finest works and creations. It's important to recognize who these players were and how they interacted with each other. Each one listed here was a valuable contributor to the evolution of the Ford Mustang. Unfortunately, many of these luminaries have passed on, taking with them thoughts, memories and recollections of the history of the Ford Mustang — memories lost forever from the automotive world.

Henry Ford II

Chairman, Ford Motor Company; Industrialist; Born: September 4, 1917; Died: September 29, 1987, in Detroit, Michigan; Role: Approved building the Ford Mustang; Father: Edsel Ford; Grandfather: Henry Ford

Henry Ford II was born and raised and died in the Detroit area. He was married to Anne McDonnell, and they had three children: Charlotte was born in 1941, Anne in 1943, Edsel in 1948. He was known for his high-pitched voice and bright, piercing baby-blue eyes, covered in later years by Ben Franklin tortoiseshell glasses. He was educated at Yale University and entered the U.S. Navy in 1941 as America entered World War II. Henry II was president of Ford Motor Company from 1945 to 1960; he was an alternate delegate to the United Nations in 1952, appointed by President Dwight Eisenhower. He became CEO and Chairman of Ford Motor Co. from 1960 to 1979; he resigned that office in 1980 after growing Ford into the fourth largest industrial corporation in the world. He brought the ten-member "Whiz Kid" team to Ford Motor Company, which revolutionized the way Ford did business. Two of those team members became Ford Motor Company presidents.

Henry II was credited with the financial rescue of Ford Motor Company after the end of World War II. His employees knew him as smart, and he showed decency and compassion for them. He was known for being able to recognize intricate factory equipment and appreciated what it did and how it did it. Considered to be always up to date on current trends, he was aware of competitive products from General Motors and Chrysler. He could talk color and shape, could talk to people about tools, and could relate to protégé Robert McNamara about figures and production volumes. He loved his company and it has been said in Detroit vernacular that "he had gas in his veins." Stunned by the failure of the Edsel project, he barely approved a new model called Mustang in 1962. He loved manual transmission cars, one of which was a '68 Cougar four-speed. Ford line manager Lee Moskowski reported all cars ordered by Henry were black with black interiors.

During the '70s he led the industry by hiring a large number of African Americans at Ford and was active in community pro-black labor organizations, for which he was well respected. "He reproduced Henry Ford Sr.'s compartmentalized, personalized authority structure and unleashed the force of his will," as reported in *The Fords* by Peter Collier and David Horowitz. He became a man of enormous power. In his widely known views on the legal profession, he reportedly disliked lawyers because he saw them as costly, time-wasting know-it-alls. He fired Lee Iacocca in 1978.

Robert Strange McNamara

Born: September 9, 1916; Died: July 5, 2009, in Washington, D.C.; Role: Considered the father of the Ford Falcon

Robert McNamara was born in San Francisco, California, and married his childhood sweetheart Margaret in 1940. Between 1940 and 1945, he and his wife both contracted mild cases of polio. He had two daughters and a son; his wife died of cancer in '81, and he married Diana in 2004.

He was an Eagle Scout, has a B.A. in economics from the University of California at Berkeley and an MBA from Harvard '39. He served in the U.S. Army from 1943 to 1946 and attained the rank of lieutenant colonel. He did statistical analysis of the bombing efficiency and effectiveness of B-29s commanded by noted Major General Curtis LeMay. He was offered a full professorship at Harvard at the end of the war and would have accepted, but he had medical bills for his wife he could never pay on a professor's salary. McNamara went reluctantly with the "Whiz Kids" to Ford in '46 and was second in command. As one observer put it, "He would have been just as happy selling widgets or ladies' undergarments." He was calculating and cold, cared only for the bottom line, and was known as a bean counter. Known as extremely smart, he was very conservative in his thinking. The Whiz Kids helped the company stop its losses and administrative chaos by implementing modern planning, organization, and management control systems. As for his physical appearance, he is described in *The Fords* by Collier and Horowitz as having "hair slicked down and precisely parted as if by calipers, granny glasses like impenetrable discs, [a] bulldog jaw — all suggesting pugnacious negation." He singlehandedly stopped production of the two-seat 1957 Thunderbird in favor of a 1958 four-seat version. McNamara rose to power at Ford too late to stop the '57–'59 retractable steel Ford hardtops, but did end them after '59. He was instrumental in the demise of the Edsel line of cars, almost closed the Lincoln Division after the dreadful '58–'60 group of cars, and caused the reinvention of Lincoln styling for the '61 Lincoln, which became very successful. His greatest achievement at Ford was the introduction and production of the top-rated Falcon car in late '59, which he conceived as a people's car, much the same concept as the Volkswagen. Lee Iacocca called him a good teacher and said, "I learned a lot from him." Design manager Eugene Bourdinat said, "He was a concise speaker and logician. He tried to apply logic to everything, and frankly, in product, you're not trying to tell the person the car he must have, you're trying to find out the car he wants, and, in that, there is an awful lot of emotionalism. In fact, the design business is probably 40 percent logic and 60 percent emotion, whereas McNamara was 100 percent logic and zero emotion." In 1960 he became the first president of Ford Motor Company selected from outside the Ford family. Ford cars of the early '60s were described as like McNamara himself, with rimless glasses and hair parted down the middle. Conservative in his ways, he had a major positive influence on Ford Motor Company spread over many years and changed the automotive design philosophy and internal operations of the company. His brilliance influenced Ford operations years after he resigned. McNamara left Ford and became President Kennedy's Secretary of Defense from 1960 to 1968 during the Vietnam War. He was president of the World Bank from 1968 to 1981. Upon his death, Lee Iacocca said, "Bob was a visionary. In the '50s he was pushing for small cars, safety features and the environment. How different history would be had he remained as Ford's president instead of joining JFK's team as Secretary of Defense."

In early 1961 there would be an informally organized group of Ford executives called together by Ford Division General Manager Lee Iacocca, which became known as the Fairlane Committee. This group would embrace the first real organized thoughts about research and product development, leading to the creation of a new Detroit concept in car building that would take the shape of a four-seat sports car.

Lido Anthony "Lee" Iacocca

Born: October 15, 1924; Age: 85 in 2009; Role: Fairlane Committee member, President of Ford Motor Co.; Referred to as the father of the Ford Mustang

Lido (Lee) Iacocca was born in Allentown, Pennsylvania, of Italian immigrant parents. He married Mary McCleary in 1956, and they had two daughters, Kathryn in '59 and Lia in '64. Mary died in '84 from diabetes. He currently resides in Bel Air, California.

His educational degrees are: Lehigh University, B.S. in industrial engineering, '45; Princeton, M.S. in mechanical engineering, '46; honorary doctorate degrees from Duke University, University of Michigan, Michigan State University, George Washington University, Fairleigh Dickinson University, La Salle University, Hillsdale College, Johns Hopkins University, and the University of Southern California. He is honorary trustee of the Iacocca Institute at Lehigh University; chairman emeritus of the Statue of Liberty-Ellis Island Foundation; chairman of the Advisory Board of Nourish the Children Foundation; chairman of the Corporate Support Committee of the Joslin Diabetes Foundation; and chairman of the Iacocca Foundation. He became president of Ford Motor Company in 1970 and was fired from Ford Motor Company by Henry Ford II in 1978 after 32 years at Ford. He became CEO and chairman of Chrysler Corporation in '78 and retired as a director in '93.

As a kid, he developed a keen interest in autos from his dad, who owned a Model T car rental business. His first job was with a small Ford dealer in Pennsylvania. The year he entered high school he contracted rheumatic fever. This made him ineligible for U.S. military service during World War II. Iacocca went to work for Ford Motor Company in 1946. He kept little black books to plan and track his career. In one of those he wrote his desire to become a Ford vice-president by age 35. Eighteen days after his 35th birthday, Henry Ford II promoted him to VP. One year later he replaced Robert McNamara as Ford Division general manager in 1960. He came to power too late to have much influence on '61 or '62 model cars, but had great influence on the "Performance Era" 1963s, which resurrected Ford Motor Company involvement in auto racing. He created and led the Fairlane Committee, which developed the new concept for a car to become known as the Mustang; he then sold the concept to Henry Ford II and saw the car through to production. Due to Mustang's success, Henry II promoted him in January 1965, to vice-president of the Corporate Car and Truck Group. He was instrumental in creation of the Mercury Cougar and the Lincoln MK III in 1968, the first Lincoln to produce a profit for Ford.

One evening in the 1970s, Henry and his wife Cristina were visiting the Iacocca household for an Italian dinner prepared by Iacocca, who considered himself an accomplished Italian cuisine cook. Henry stated to Iacocca's father Nicola, "I don't know how I could run this company without your son." Later, in 1978, Iacocca was fired by Chairman Ford after the biggest single moneymaking month and twenty-four-month period in Ford history. He became chairman of Chrysler Corporation and is best known there for creating the very successful front-wheel drive minivan, which is still being produced. Iacocca is credited with rescuing Chrysler from bankruptcy in the late '70s, and served as a director until 1993.

He was asked by President Ronald Reagan to chair the Statue of Liberty–Ellis Island Foundation in '82, established to renovate and restore the Statue of Liberty. He raised over one-half billion dollars for the project. Lee Iacocca has been on the cover of *Time* magazine four times, compared to only three times for Henry Ford II. In a 1987 Gallup Poll, he was ranked second on the list of the world's most admired people, behind Ronald Reagan and ahead of Pope John Paul II, and was the first businessperson ever named to the poll's top 10 list, beating out even Henry Ford I. He has written three best-selling books. He was asked to run for United States president in the early 1980s, but declined. In 1984 he founded the nonprofit Iacocca Foundation to help find a cure for diabetes, whose funded research efforts continue today.

Dr. Donald Nelson Frey

Born: 1923, St. Louis, Missouri; Died: March 5, 2010, in Chicago, Illinois; Role: Fairlane Committee member, Ford Product Manager, Mustang Project Manager, co-originator of the sporty two-seat concept that became the Mustang Experimental Prototype

Dr. Donald Frey was married to opera soprano Helen-Kay and they had six children. His last residence was in Chicago, Illinois.

He was a World War II Army officer from 1942 to 1946. He received a PhD from the University of Michigan in '50. Frey spoke Russian and French. He was hired by Ford in '50 and retired in '68 as Ford Division president. He became chairman of Bell & Howell Corporation, where he was instrumental in the invention of the first CD-ROM. In 1988 he became a professor at Northwestern University in Chicago. He was awarded the prestigious National Medal of Technology in '90 by President George H.W. Bush. He has been called an innovator in manufacturing and informational systems and owned a 1964½ Mustang hardtop till death. Frey liaisoned with the Budd Company to create a two-seat design prototype called the XT-Bird based on the '57 Thunderbird early in the concept stages of Mustang development. Frey was responsible for the first disc brakes on an American production car from the Big Three ('65 Mustang, '65 Thunderbird) and for introducing Ford to the radial tire first used by Ford on the Thunderbird and Lincoln. Frey insisted on a second turn signal light on the '65 Mustang dashboard after dealers said the one light looked too cheap. He championed the idea of the Bronco 4 × 4 truck built in '66. Frey was given a red early production '65 Mustang coupe, highly modified by the styling department, to drive; its present whereabouts are unknown. His two heroes in the Mustang development project were Joe Oros, "who introduced Italian styling to the car," and Hal Sperlich, "who figured a way to keep costs low by utilizing the existing Falcon chassis." Frey was considered by all to be a gifted engineer. Although in the Mustang project he was always there, he didn't have day-to-day hands-on involvement as did Sperlich and Iacocca, and did not provide a lot of the decision making. Regarding parts and material supplier gifting to Ford employees, he stated that rule #1 at Ford when he took control was "You can't take anything more than you can use or eat in a day," a flagrant practice which he stopped.

Harold K. Sperlich

Born: 1930; Age: 79 in 2009; Role: Fairlane Committee member, Special Projects Assistant to Don Frey, Mustang Program Manager

Harold (Hal) Sperlich and his wife Polly live in Orchard Lake, Michigan. They raised a family of seven children.

Sperlich received an engineering degree from the University of Michigan in 1951 and then an MBA degree. It was Sperlich who came up with the brilliant idea of using the Falcon platform and body parts for the production Mustangs. He is often referred to as the true father of the Mustang because of his idea of building a new class of car using an existing chassis. Without this creative idea, the Mustang may have never been built due to the limited budget Henry Ford II gave the committee to create the car. As an engineer, he was one of the first in the industry to push efficiency and smaller car design. Stylist John Najjar called him "dynamic, very creative, with an alert mind and agile with figures. He understood the marketplace and the need for good vehicles out there. That is why Iacocca had him on the side under his wing. He was always in there pitching." He was considered by some as "too pushy, too bold, too abrasive, too outspoken, but usually 100 percent right." One designer said, "Sperlich was a smartass. He and Iacocca used to sit around and ridicule Henry behind his back." Henry never liked Sperlich, saying he was too argumentative. According to Collier and Horowitz in *The Fords*, in '77 Henry asked Iacocca to fire him. Iacocca responded, "You've got to be kidding; he's the best we've got." Henry said, "Fire him now ... if you don't can him now, you'll go out the door with him. And don't give me any bullshit. I don't like him and you're not entitled to ask why!" Sperlich was fired. He was almost immediately hired by Chrysler for his brilliance in design innovation. After Iacocca became CEO of Chrysler, he used Sperlich as one of the key members of his reorganization team. He was an original proponent of front-wheel drive cars, believing them to be superior. He developed the minivan concept using the Chrysler-subsidiary-built Simca car front-wheel drive and the Chrysler K-car platform with a firm belief Americans would buy them. Iacocca states, "Twenty years after the Mustang, baby boomer families grew, they wanted utility in their cars. So Sperlich and I hit them again with the minivan, which is still in production today at 40,000 units per month!" Sperlich said, "If Henry Ford II had accepted my minivan concept, Chrysler would not exist. The minivan is the backbone of Chrysler today...." He was correct, and these cars carried Chrysler out of bankruptcy. He was made president of Chrysler in '85 and remained a confidant to Iacocca. He left Chrysler in '88 for a short retirement and in '94 became part owner with General Motors of Delco Remy America, a parts supplier. He was inducted into the Mustang Club of America Hall of Fame in '98. Edsel Ford II said in 1989, "Sperlich is one of the greatest product people ever to grace our industry. His ability to envision what consumers want was phenomenal." Iacocca said, "Harold was the product genius at Ford and Chrysler. He was the chief architect of the two most important products in my career: the Mustang and the Minivan."

Sperlich is quoted: "There's very little accomplished by one man sitting at his desk having brilliant ideas. Good ideas generally come out of an interaction with many people and it was that way with the Mustang."

Frank E. Zimmerman, Jr.

Born: October 10, 1924; Died: March 2006 in South Carolina; Role: Fairlane Committee member, Ford Marketing

Frank Zimmerman was a fellow classmate with Lee Iacocca in a "Ford Introductory Course" when hired at Ford. He would become a good personal friend to Iacocca, who would call him "Zimmie." Iacocca said he was the "resident promotion genius and he worked well with Hal Sperlich." He was thin as a reed, had endless energy and was very funny. He was reported as a joy to work with and was an unforgettable character with a new idea every 10 minutes. Iacocca chose him to work with the committee because of his creative mind. He interacted with Sid Olson to develop the original Mustang advertising. It was Zimmerman who came up with the idea later of using a live cougar in the 1967 Mercury Cougar advertisements and he developed the successful phrase "At the sign of the Cat," which Mercury used for years.

Walter T. Murphy

Born: 1916; Died: February 19, 2002, in Beverly Hills, California; Role: Fairlane Committee member. Ford Public Relations Manager

Walter Murphy was married and had two daughters and three sons.

He attended college and was a captain in the Army Air Corps during World War II. He joined Ford Motor Company in 1947. Known as a quiet, pipe-smoking man, Murphy headed the Ford public relations team for 32 years with 400 executives working for him worldwide. He spent 20 of those years personally as a public relations advisor, speech writer, corporate confidant and press agent for Ford company presidents. Walter Murphy instituted the first Mustang advertising program and was known as "Iacocca's man." Iacocca told me, "He was an excellent public relations guy!" He chose to retire in 1978 rather than accept Iacocca's invitation to join him at Chrysler. Lee Iacocca reports in his book *Iacocca*, "On July 13, 1978, the day I was fired by Henry Ford, Murphy worked in his office until midnight, and was in bed by 1 A.M. The phone rang by his bed at 2 A.M. and the conversation went like this: 'This is Henry.' 'Henry, Henry who?' 'Henry Ford.' 'Yes, Mr. Ford.' 'Do you still love Iacocca?' 'Do I what?' 'Do you still love Iacocca?' 'I like him very much, of course.' 'Then you're fired too. Starting tomorrow.'" Henry rescinded the order the next day, but Murphy soon after retired.

Sidney (Sid) Olson

Born: September 8, 1922; Died: November 27, 1986, in Pennsylvania; Role: Fairlane Committee member, represented J. Walter Thompson Advertising, Ford's advertising agency

Sidney Olson was born on an Indian reservation near Tulsa, Oklahoma. He and Patricia were married in 1950. He advanced to corporal in the Army and worked on Army newspapers. With no formal education, he joined the advertising firm J. Walter Thompson Company in 1947 and wrote advertising copy for the Ford account from 1953 to 1971. He was once a speech writer for President Franklin Roosevelt. Olson coined the phrase "the arsenal of democracy." He was described as a "brilliant advertising copywriter" and editor and was involved in coordinating the initial advertising campaign for the new Mustang.

Robert J. Eggert, Sr.

Born: December 11, 1913, Little Rock, Arkansas; Died: November 18, 2007; Role: Fairlane Committee member, Ford Market Research Manager

Robert Eggert and his wife Betty raised three sons together.

He received BS and MS degrees from the University of Illinois, and a PhD from the University of Minnesota. He worked on the development of early Thunderbirds and the Falcon. He primarily did marketing and economic research at Ford and is the one reportedly who first suggested the name "Mustang" at a Fairlane Committee meeting, and said, "It came from a book my wife gave me for Christmas in 1960." He stated they tested 17 different names for the car before deciding on Mustang. Later in retirement he founded the Blue Chip Economic Indicators Newsletter, taught at 5 different universities, and became an economic advisor to the U.S. Congressional Budget Committee. He retired in Sedona, Arizona.

Chase Morsey, Jr.

Born: December 11, 1919, St. Louis, Missouri; Age: 90 in 2009; Role: Fairlane Committee member, Ford Division General Marketing Manager

Chase Morsey is married to Beverly, with a family of three sons and three daughters.

As a young man, Chase Morsey joined Ford in 1947 as a trainee. In a department training course, he was given a book to read about the yet to be introduced '47 Ford, which mentioned only a 6-cylinder engine to be available because it cost $100 less than a V-8. According to a *Fords of the Fifties* notebook article of December 2000, he told his boss, "This will break the company. The V-8 *is* the Ford car. I'm a Ford guy and the V-8 is what makes guys buy Fords." So he went with his instructor to see manager Lewis Crusoe about it. They agreed to do a survey with Ford dealers about keeping the V-8 in the '49 car. A consensus of 90 percent said keep the V-8, even though competitor Chevrolet offered only a six-cylinder. The decision was then made to keep the V-8 in '49. This was a good decision, Morsey reported to me in April 2008, "because it sold an additional 200,000 '49 Fords. This led the way for designing the new body style '52 Ford with an optional V-8. Had only a six-cylinder been designed for the car, it would have been a major setback to Ford sales in to the '50s." He was part of the Competitive Analysis Department in the early '50s, which involved clandestine spying on General Motors. Morsey told me he and others were sent to Arizona to climb fences at the GM facility with cameras strapped to their necks to spy on the new designs of the '50s. At Ford, he saw a proposed '56 Continental MK II roofline in the secret Continental styling office and took the idea back to the Ford Division, where they "borrowed" the idea of using that roofline with the blind quarter on the 1955 Thunderbird. He was the head of market research for the Fairlane Committee. It was his research that showed the demographic group called "baby boomers" was just reaching car-buying age in the early 1960s. The results of his research would have a profound effect not only on Ford, but on the whole industry's thinking during the '60s. Later in the '60s, he worked for RCA as the chief financial officer. Morsey was a quiet individual with an aggressive personality and was superior in his knowledge of the advertising business.

Jacque Passino

Born: 1920; Living in New Bern, North Carolina, as of 2008; Role: Fairlane Committee member, became Ford Special Vehicle Division Manager and Racing Director

Jacque Passino is married to Floranne. Their family includes one daughter and three sons. Passino has a mechanical engineering degree received from the University of Toledo in Toledo, Ohio.

From 1959 through 1963, Passino kept the pressure on Ford management to resume involvement with factory-sponsored racing. He was responsible for supporting the Holman Moody Racing team program sponsored by Ford in the late '50s and '60s. Under his direction the Ford return to NASCAR and Indy-style racing was formulated. He toured the United States with the original two-seat Mustang, showing it at many colleges and local shows. In 1961 he began the Ford Indy 500 race engine program. In 1968, he convinced Ford to put the big block 428CJ engine in the Mustang. His nickname was the "Gray Fox."

John Bowers

Born: unknown; Died: unknown — reported deceased; Role: Fairlane Committee member, Ford Division Advertising Manager

Family information is unavailable.

No facts are known about Bowers's education.

John Bowers was hired by Chase Morsey Jr. to work in his department at Ford as a Ford Division advertising manager. Research done by Bowers was brought to the Fairlane committee and influenced marketing objectives and plans for the new car. He was last reported living in Grosse Point, Michigan. It is reported he was proud to always wear his Phi Beta Kappa pin to work. Jacque Passino called him "a nifty guy. He was on the quiet side but had a very forceful personality. He was always very clever with the ideas he brought to the committee."

Although the following personalities were not members of the first-tier Fairlane Committee as such, they were all instrumental in the design and creation of many Ford vehicles, including the Mustang, during the time the Mustang car was in its developmental stages.

Eugene Bordinat, Jr.

Born: February, 1920 in Toledo, Ohio; Died: August 1987 in Detroit, Michigan; Role: Project Manager for the 1962 two-seat Mustang prototype

Eugene Bordinat was married many years to Teresa.

He was raised in a Detroit suburb where his father was an engineer at Willys-Overland and Chrysler. He studied art at Cranbrook School in Michigan (two years). Bordinat attended the University of Michigan until '39. His first job at GM in '39 was working on styling of the '41 La Salle and '42 Chevrolet. He was in the Army Air Corps for two years in World War II. He was hired by the Ford Design Department in '47. He became vice-president and general manager of the Ford Design Studio in '61 and retired as a Ford Motor Company vice-president in '80. Bordinat worked earlier as a stylist for Harley Earl at General Motors. He developed the very distinctive side design and the first Ford vinyl roof on the '50 Ford Crestliner, designs he borrowed from Duesenberg cars. When requested by the Fairlane Committee, he attended the original committee meetings as a design consultant. He coordinated styling of the body of the 1962 two-seat Mustang from drawing to clay in just three weeks. He oversaw building of that Mustang and used Roy Lunn as product planner and Herb Misch as project engineer, and together these three were credited as the "magic" behind the two-seat Mustang project. Bordinat oversaw styling of the 1965 Mustang, '60 Falcon, '55 Lincoln and the Lincoln Town Car. He was good friends with Joe Oros and was known as "Frenchie" to some of his close friends, and "the Frenchman" to Iacocca, who called him "a very dapper, smart dresser and a great stylist, fun business associate and he ran a good shop." He supported continuing the Edsel line of cars until its demise. He teamed up with designer Hal Sperlich to sell the idea of the minivan to Iacocca, and influenced the design of 50 million cars and trucks in the U.S. alone. He served under eight Ford presidents. Henry II once told him, "Gene, you're the biggest brown-nose I've ever met in my life." Bordinat responded, "Well I'm glad to hear that Mr. Ford, because anything I do, I want to do well," and Ford laughed. "Now if Ford hadn't laughed," Bordinat recalled, "I would have been in deep shit, but he did laugh." Bordinat was a bit flamboyant and in later years could be seen driving his Clenet or Mangusta, dressed with a gold pendant swinging over a scarlet ascot and with gold wrist ornaments weighing about a pound each. His quote: "The bumper's only use is a Braille device to help you park."

Herbert L. Misch

Born: 1918 in Sandusky, Ohio; Died: June 2003 in Royal Oak, Michigan; Role: Project Engineer for the 1962 two-seat Mustang. Provided the expertise for engine/drive train development

Herb Misch was married with at least one daughter.

He graduated from the University of Michigan with a BS in engineering '41. He became vice-president of engineering and research for Ford Motor Company in '62. Misch was an inventor/innovator, and worked as an engineer for Packard in the late '40s. There he was instrumental in developing the first automatic transmission in the industry which used a torque converter, called Ultramatic Shift, used in the Packard cars. He was chosen as project engineer to develop the two-seat Mustang prototype in the spring of '62. Misch commented in the October 7, 1962, Ford press release on the two-seat Mustang: "We know the sports car field is growing. Americans are taking more personal interest in their cars. They want personal cars they can feel and enjoy driving. Our advanced engineering group wanted to obtain more experience in this field, so the Mustang project was given the go-ahead." He later helped in research and development of the first safety air bag systems at Ford and was active in retirement with emissions and environmental quality research.

Royston G. Lunn

Born: 1925 in England, living near Sarasota, Florida, in 2009; Role: Executive Research Engineer, 1962 two-seat Mustang Project

Royston Lunn was educated in England, where he studied and graduated from courses in mechanical and aeronautical engineering. He began his career working as an apprentice toolmaker in England. Upon leaving the Royal Air Force in '46, he was a designer for AC cars in England. In '47 he became assistant chief designer for Aston Martin and was responsible for the legendary DB2 model program. He joined Ford of England in '53, emigrated to the U.S., settled in Dearborn in '58, and became the Advance Vehicle Center manager for Ford. Lunn was asked by Eugene Bordinat, V.P. Design, and Herbert Misch, V.P. of Engineering, in May '62 to lead the special project to design a two-seat experimental car which became the 1962 Mustang prototype. He assembled a team and worked with no budget and set a goal of having a complete drivable vehicle done in 100 days in time to run at Watkins Glen Raceway, New York, in October 1962, a goal he accomplished. Immediately after the two-seat prototype Mustang, he went on to design the GT40 race car using styling cues from that Mustang on the early GT40s.

John Najjar (pronounced Nā-jār)

Born: November 11, 1918, in Omaha, Nebraska, living in Florida in 2009; Role: Head of the design staff that created the experimental 1962 two-seat Mustang.

John Najjar was married and had two daughters. His father was Egyptian-born and his mother was Lebanese.

Known by friends as Johnny, he was hired out of high school to work in the Ford factory. While Henry Ford I was walking through the factory, he asked the young Najjar if he was happy with his job. He said he would rather be designing cars. After presenting

Henry with some of his sketches as requested, he was transferred to the Engineering Department as a drafting trainee. In the early '50s, the then-jokester placed a cherry bomb–laden three-foot rocket in his boss's desk; it exploded, causing disruption to the entire styling department. He was in the Ford design department for 37 years and was most influential on the design of '57 and '58 Lincolns, '61 Continentals, and the '61 Thunderbird. Eugene Bordinat, in his 1985 Interview for the Benson Ford Research Center Oral History Project, described Najjar as "highly intelligent, had a beautifully organized mind, not the finest designer in the world, but a good recognizer of design, he was an atrocious renderer, but an absolutely brilliant sketch man...." Najjar supervised the design of the '62 two-seat Mustang, and originated design sketches for the 1963 Mustang II. In February 1962, he was the executive stylist in charge of the Advanced and Pre-production Vehicles Studio and had posted several of his side-view drawings of a possible two-seat sports car on the wall in his studio. V.P. Bordinat showed the drawings to Lee Iacocca. In three weeks he had a clay mockup ready of the two-seat Mustang. He was in charge of interior design for the 1965 model Mustang and was the originator of the vinyl grain imprinted metal used on the interior sides of the doors as well as the curved rear folding seat on the later 2+2 model. He used the '64 Thunderbird seat design to style the new '65 Mustang seats. Although he suggested the '62 two-seat prototype be named after the World War II P-51 Mustang fighter plane, his boss rejected the airplane connotation and instead chose the name Mustang after Najjar changed its meaning to the equestrian connotation. As a manager on that project, he was associated with the design of the first Mustang emblem that was used only on that prototype, but did not design the emblem himself. He retired in 1979.

Joseph Oros

Born: 1917 in Detroit, Michigan; resides in Santa Barbara, California, in 2009; Role: Overall design responsibility for the 1965 Mustang

Oros was born of Romanian immigrant parents. Married to Betty Thatcher (deceased), a designer for Hudson and reputed to be the first woman designer for a major auto manufacturer. Together they raised five children.

Oros graduated from the Cleveland Institute of Art in 1939. Always of slight build, he came from a poor Detroit family that had no car when he was a child. After his graduation from art school in 1939, he was employed by General Motors and out of necessity learned to drive a car. He worked in the engineering department at GM where Army tanks were designed. He left GM after the war and went to work for George Walker Industrial Design Company, where he quickly was recognized by George Walker as a talented designer. When Walker was awarded the design contract of the '49 Ford car by Henry Ford II, Oros became heavily involved in the design of the car and was specifically credited with the design of the grille, which was meant to make a distinct statement that the new postwar designs were here and this was Henry Ford II's first new car. He worked as a part-time consultant to Ford on the design of the '52 Ford and was responsible for the "round taillight" concept that continued in Ford design for many years. He remained a designer with the Walker Company until '55, when George Walker was hired full-time by Henry Ford II and brought Oros with him to Ford. He is credited with the design of the '57 Ford car. It has been said that his ability to know exactly what Henry Ford wanted in the design of a car was uncanny. He also did major design work on the then-new unibody style 4-passenger '58 Thunderbird. John Najjar described him "as a good designer who had a feeling for themes. He always treated his job with respect." Other contemporaries said "he had Ford cars in his blood" and he was considered by most as a "real star" at Ford. On '65 Mustang styling, he said, "We weren't going to do anything with the styling of the Mustang that had previously been done." *Ward's Auto World* interviewed Walter Murphy in 1986 where he quoted Oros's statement, "When Lee Iacocca saw our finished car, he just rolled his cigar in his mouth. I could see the gleam in his eye, and he was pleased as punch. Of course, that made me feel very good too." Of the final Mustang design, he said Henry Ford had only one criticism of the car: he wanted more rear seat space. Mr. Oros and his design team received the prestigious Industrial Design Society's Design Award for the design of the 1965 Mustang. He became director of Exterior Design for Ford and oversaw the building of the Torino and later models of the Mustang. Oros retired in 1981.

L. David Ash

Born: April 15, 1921, in Detroit; Died: July 2, 1991; Role: Executive Assistant to designer Joe Oros in the Ford Studio, credited with major exterior styling and design work on the 1965 Mustang

Dave Ash was married to Rosemary with a family of one son and two daughters.

After high school, Ash attended Meinzinger's Art Academy in Detroit, Michigan. Ash worked initially as a designer at General Motors. He then was persuaded to come to work at Ford by Eugene Bordinat in October 1947. Ash was credited with the design of the Ford crest on the '49 Ford. George Walker used to send him out to buy expensive cigars for him at design meetings when Walker was a consultant to Ford in '49. He worked on the design of the '54–'56 glass-top Skyliner models and his first full-scale clay mockup was of the '55 Crown Victoria model. During his career at Ford, he worked on many concept and show cars for the Ford, Mercury, Lincoln and Edsel studios. He designed the interiors on the '58–'60 Continentals and also the interior of the new body style '61 Lincoln. He worked on the interior design of the '61 Presidential Lincoln, the car in which President John Kennedy was assassinated. While designing that car interior, he questioned the chief of the White House Secret Service staff and recommended the car have a bulletproof top, which the Secret Service disapproved (according to the L. David Ash 1985 interview for the Oral History Project of the Benson Ford Research Center). Other design projects he is credited with include the GT40 race car and the '69 Lincoln Mark III (specifically at the direction of Lee Iacocca). In the early '60s he was transferred exclusively to the Ford design studio, where he was associated with the interior design of the '64 full-size Ford and Fairlane models. He was assigned to the Oros design group, where he was credited with the design of the side scoop on the '65 Mustang, an idea he said he got from one of the '58–'60 Continental prototypes that was never produced. He later assisted in design of the '67 Cougar and was elevated to chief stylist and manager of the Ford Exterior Design Studio by Lee Iacocca for his design work on the Mustang. Iacocca called him one sleepless night and told him to design a new Lincoln with a Continental tire design on the trunk lid, to be called the Merlin. Ash produced a design that became the 1969 Mark III Continental, which was named by Iacocca. Ash said he bought and drove an Edsel hardtop, which he liked very much, for many years. He retired from Ford in 1981.

Gale L. Halderman

Born: 1932; living in Dearborn, Michigan, in 2009; Role: Member of Joe Oros design team in the Ford Studio, credited with the exterior design of the 1965 Mustang

The Haldermans, Gale and Barbara, still reside in Dearborn, where they raised four daughters.

Gale Halderman grew up in Dayton, Ohio, and graduated from Dayton Art Institute in 1954. One of his classmates was Jonathan Winters, the famous comedian. In 1955 he was hired by Eugene Bordinat as a designer in the Lincoln-Mercury Studio. He was transferred almost immediately to the Ford Design Studio, where he began design work on the '57 Ford, including the new retractable hardtop model. Halderman is given credit for major styling of the 1960 Falcon. He also worked on the design of the new model 1961 Thunderbird. He attended some of the original Fairlane Committee meetings as a design consultant at the request of the committee. Halderman worked for Joe Oros with Dave Ash designing the '65 Mustang. These three are credited with the major styling and design of the new Mustang. In my October 2007 interview he stated, "Over a weekend I produced 5 styling sketches on my kitchen table [at] 10 P.M. at night. Oros liked one of the sketches in particular, and the car styling evolved from there, with the first clay model originating from that sketch." Oros credits Halderman not only for contributing to the design but also for skillfully guiding the Mustang from clay-model dream to a realistic, fully producible car. He worked on the design of most Mustangs through the late '60s. He became the head of the Ford Design Studio and in 1968, head of the Lincoln-Mercury Design Studio. Halderman later declined three separate invitations from Iacocca and Sperlich to work at Chrysler. He retired in 1994.

Robert D. Negstad

Born: March 26, 1930, in Portland, Oregon; Died: January 2001 in Dearborn, Michigan; Role: Project Engineer for Chassis Design, 1962 Mustang, 1965 Mustang

Bob Negstad was married and had one stepdaughter, Nancy.

Born to working-class parents during the depression, Negstad found interest in working on cars in high school and in college at Oregon State University. His college education was interrupted when he dropped out of school to fight in the Korean War. He served in the Army from 1951 to 1953. Upon returning to Oregon State, he graduated in 1957 with a BS degree in mechanical engineering. He was recruited into the Ford Motor Company college apprentice training program and worked in Dearborn during summer break. His dream was to work for Henry Ford, and in 1956 he drove his '57 T-Bird to Detroit, where he started a 30-year career. He worked in the Car Product Research Department, specializing in advanced chassis and suspension design. He was closest of friends with racing legend Jack Roush and they both showed their own classic cars together at many car shows up until his death. He worked for Roy Lunn and designed the chassis and suspension for the 1962 Mustang at the age of 32. One of the drivers of the Mustang at its debut at Watkins Glen Raceway in New York was Bob Negstad. He was also one of the staff that took the two-seat Mustang on tour around the U.S. in 1962–63 to display the prototype at various colleges. He introduced Ford to the front-wheel drive concept which was later developed by Hal Sperlich. He redesigned the suspension for the Shelby Cobra for Carroll Shelby, designed the suspension for the GT40 race cars and participated with car support at LeMans in 1965. He designed the suspension for the Shelby GT350 and the Shelby 427 Cobra. He was transferred to the Special Vehicles Operations department at Ford in 1980 and was lead project engineer for one of the first SVO cars, the 1984 SVO Mustang. He died in 2001. It was the author's distinct honor to personally know Bob Negstad. He assisted me in authenticating my First Production Mustang Hardtop, VIN 5F07U100002, in the late '90s.

George W. Walker

Born: May 22, 1896, in Chicago; Died: January 19, 1993, in Tucson, Arizona; Role: Initiated first design drawings of 1955 Thunderbird, brought Joe Oros to Ford Motor Company

Walker was married to Frieda with two children.

George Walker, although not directly related to the conceptualization of the Mustang, was a major player in the recovery of Ford Motor Company after World War II. He declared himself, along with industrial designers Harley Earl and Raymond Lowey, as part of the "Triumvirate" of modern industrial design, stating, "We did more for the industry than anybody could have done." He was considered the "styling chief with star status." He brought his employee Joe Oros with him to Ford in 1955. His condensed biography conveys a feel for the atmosphere that permeated the Ford styling department of the late '40s and '50s.

He attended the Cleveland School of Art in Cleveland, Ohio, and the Otis Art Institute of Parsons School of Design in Los Angeles. He began his career designing women's clothes for magazine ads placed in *Vogue* and *Harper's Bazaar* magazines. A semi-pro football player in the '20s, he opened his own industrial design firm in the late '20s. He did design work for Pierce-Arrow, Packard, Willys, Graham-Paige, Nash and International Harvester. He was made a consultant to Ford in '46 and his firm was called upon by Henry Ford II to design the '49 Ford, whose popular car sales were credited with saving Ford. When he first saw the Ford in-house styling design of the '49 Ford, he said to Henry Ford II, "It looks like me stooping over an oven and it looks kind of fat. You should make it a Mercury." And that's what they did with that design. He redesigned the '49 Ford on a crash basis with the help of his employee, Joe Oros. After 1949 they both left Ford, but advised in styling the '50 Lincoln, the '51 Mercury and the '52 Ford. He then returned in 1955 with Joe Oros and stylist Elwood Engel, all as full-time employees, after years of consulting for Ford. His contempt for underlings could be seen at design meetings when he would send Dave Ash and John Najjar out in the middle of the meeting to buy expensive cigars for himself. Henry II later made him a V.P. based on his involvement and the success of the '55 Thunderbird. He always said, "Lee Iacocca had no style." He had a back-slapping manner that charmed men and women alike and was a memorably blunt man, tough-talking and crude when he wanted to make his point. He was considered a political animal. Walker was called "the Cellini of Chrome" by *Time* magazine and was featured on a cover of *Time*. He bragged in the *Time* article, "A stylist had to show style in his cars, in his house, his clothes and his person." Walker was equal to the public demands made on corporate styling chiefs. His vast office was carpeted in soft mouton, his large desk was kidney-shaped. A semicircular banquette and low, round coffee table provided meeting space. Soft music added to the atmosphere. His 70 suits were custom made; his cologne, lavishly applied, was by Faberge. He owned 40 pairs of shoes. His flamboyance is perhaps best summed up by Walker himself, recounting what he considered his "finest moment" on a Florida vacation: "I was terrific. There I was in my white Continental, and I was wearing a pure-silk, pure-white embroidered cowboy shirt and black gabardine trousers. Beside me in the car was my jet-black Great Dane... . You just can't do any better than that." He knew everyone in the auto business. Walker helped design 3000 products through his industrial design business, including watches, radios, washers, alarm clocks, etc. He was appointed vice-president of design and general manager of the Ford Styling Center on May 22, 1955. He always encouraged modernistic and fancy concept cars, and usually instigated conflict and competition between designers and modelers to create new ideas. He always bragged he went to school with Bob Hope's brother. Most importantly to this story, he is the one who brought Mustang stylist Joe Oros to Ford. Walker was required to retire in 1961. He moved to Florida, where he was mayor of the city of Delray Beach.

Phillip T. Clark

Born: October 27, 1935; Died: February 28, 1968, in Nashville, Tennessee; Role: Designed the 1962 Mustang horse-and-bars emblem similar to all subsequent production Mustangs, and was credited with some vehicle styling of the '62 two-seat Mustang prototype.

Phillip Clark was married to Marilyn and they raised two children.

Clark had thoughts of designing a line of sporty two-seat cars around a horse theme while in high school. He sent his sketches to Chrysler Corporation, who rejected them, but recommended he attend a school of design to further his education. After attending the Art Center School in Pasadena, California, he went to work for General Motors in 1961. There he assisted in designing the "Car of the Future," which was eventually displayed at the 1964 World's Fair GM Futurama display. Clark introduced his conceived "Pony" two-seat design theme with vehicle and emblem sketches to Ed Mitchell, the GM styling chief. That theme was rejected and Clark left GM in early 1962. According to tax records recently uncovered from Clark's estate by his daughter Holly, he joined Ford in early spring '62. He carried his preliminary "pony" design and emblem sketches with him to the Ford Design Studio, where the new concept two-seater was in an early design stage.

A short time after Clark left GM for Ford, and as reported by Gary Witzenburg in his book *Mustang*, Chuck Jordan at GM Styling called to ask John Najjar if they were working on a new car called Mustang. Najjar confirmed that they were and Jordan is quoted as saying, "Damn, we just finished a special vehicle for GM styling chief Mitchell they named Mustang—with the horse and everything on it and here you guys do it. Don't you have any other names?" Clark's artistic talents were applied to the "pony" running horse emblem to fit styling requirements for the 1962 two-seat Mustang. Boss John Najjar said of the emblem design, "It was a natural because the hard vertical straight lines counterbalanced the movement and fluidity of the horse! Red, white and blue gave contrast to the emblem. We made bas-relief cardboard drawings of the emblem and mounted them on the model. The emblem established the identity of Mustang for the vehicle." Clark was in ill health when hired by Ford, a factor which delayed his formal hiring several months. Pending medical test results from the Mayo Clinic in Minnesota, he would receive proper credentials as a full-time employee after final plans for the '62 Mustang were completed. His involvement with the Mustang was as a consultant, not an employee. If he had not been allowed to work on that project, his emblem design work may have never been integrated into the design of the two-seater emblem. After a move to England a few years later, he returned to the United States and was terminated by Ford, presumably due to continuing health problems, according to his daughter Holly Clark. He applied for employment at Chrysler, but was unable to complete the new hire process due to ill health. Part of his resume to Chrysler included his renderings of a new design for a minivan-type vehicle. Just a few short years after design work on the Mustang emblem, Phil Clark, who subsequently worked on later Mustang and Thunderbird styling, would die in early 1968 from kidney failure at the age of 32. Jack Telnack, good friend of Clark's and later Ford V.P. of Corporate Design, wrote in a letter to Clark's daughter Holly, "If

Clark had lived, cars would be different now. His designs then were just way far ahead of his time." In an August 9, 2002, letter, J. Mays, V.P. of Design at Ford, said, "The legacy of Clark's styling was the drawing of the galloping horse used on the Mustang logo" (for the two-seat Mustang prototype).

Many of these men worked during their careers, both before and after Ford Motor Company, for other car manufacturers as designers and stylists. That was common in those days. Many of the ideas and designs for styling were gleaned from associations with other high profile stylists like Harley Earl at General Motors and Virgil Exner at Chrysler. There was definitely an interaction between these relatively few outstanding men. A homogenization of styling cues can be found on all of the cars made in Detroit during the '50s and '60s. It was not unheard of for a General Motors stylist to have a casual friendly conversation with a longtime friend who was a Ford stylist. Associations and friendships made in colleges and art schools carried new car design across the board to a new high during the '60s so-called Performance Era years. Styling cues were transferred to similarly designed cars at GM, Ford and Chrysler through the interactions of these friends. As an example, there were clay model Chevrolet design drawings secretly obtained by the Ford stylists for the purpose of trying to improve on the Chevrolet design, of course, just as a styling exercise.

Even though Detroit would have the public believe that all things are secret from each other in the world of car design, there were covert "cracks in the plaster." Welcome to the stylists' broad horizons and world of imagination.

Chapter 5

1960 Detroit Compact Revolution

"Kick the hell out of the status quo."
Ed Cole, President, General Motors Corporation

It all started in 1959, the same year rock singers Buddy Holly, the Big Bopper and Ritchie Valens died in an Iowa plane crash. For the Ford family, the decade would begin on a positive note. On January 10, Bill Ford purchased sole ownership of the Detroit Lions football team for $4.5 million. He still owns the franchise today. Immersing himself in the team would become a lifelong passion.

The nation's baby boomers had now come of age and were entering economic markets where they had had no previous influence. The decade would be framed by their growth into a political and a financially driven society. As the mid-twenty-year-olds married, purchased their first new homes and had babies, they were ready for a new breed of automobile. The wheels had been set in motion in the last years of the previous decade by the "Big Three" auto manufacturers. Detroit would give them just what they wanted: a new kind of economically viable car to fit their tight budgets. They needed something cheap, economical to operate and downsized from the late '50s behemoths then on the road. Women especially expressed in surveys that a second car for a two-car family should be a downsized smaller auto. It was found that most buyers would prefer an American-built smaller, cheaper car, as opposed to a foreign-built car like Volvo or Volkswagen.

The continued interest in sports cars suggested a new market segment for Detroit in the early '60s. Each of the "Big Three" auto manufacturers would introduce cars that were sporty and low priced.

The marketplace was stunned when General Motors, Ford and Chrysler announced almost simultaneously they would each produce a revolutionary economical small car affordable to all. These new marvels would become known as compact cars.

For General Motors, the entry would be called the Corvair. Chevrolet chief engineer Ed Cole, father of the Chevrolet small-block V-8 and designer of the '55–'57 Chevys, revived the name Corvair from the possible name selection file of the prototype Corvette in 1953. GM wanted a new, small, lightweight vehicle to compete with the imported small cars now arriving in the United States. Experimental design was begun in 1956 with the first production GM vehicles introduced October 2, 1959, as 1960 models. The so-called A-body design incorporated unit-body construction that was stronger than the separate body-and-chassis designs of the time. Ed Cole was enamored with airplanes and felt his new Corvair would benefit from an aircraft-derived engine. It would be a rear-mounted, horizontally opposed 6-cylinder air-cooled design, similar to the Volkswagen engine. The soapcake-like body using the flat engine technology resulted in what was a very different car from any other that General Motors would ever introduce.

It turned out to be the most controversial Chevrolet since the 1920s. The avant garde car was built in response to the anticipated market conditions at the time, but its radical design made it costly and foreign for the audience it was to target. Regardless, with a base price of only $1,984, the car sold well, with the appetite of the country swallowing up the first Chevy compact car with over 253,000 sales.

Corvairs were offered later as convertibles, station wagons, vans, and Loadside and Rampside pickups. The Corvair was the first of the so-called compacts to offer factory air conditioning. It had flat floors *sans* the usual transmission hump since the engine was rear mounted. When the sportier version of the base two-door car named Monza was introduced, it grabbed the bucket-seat, four-speed floor-shift audience immediately.

Other GM clone compacts followed suit with the same sporty interior modifications, including the Buick Special Skylark, the Olds F-85 Cutlass, and the competing Lark Daytona from Studebaker. They all could be ordered with bucket seats and four-speed shifts satisfying any need for a sporty interior.

Corvair was introduced with a 140-cubic-inch six-cylinder engine with 80 horsepower. The car was overweight and underpowered from the beginning. The aluminum

Early Chevrolet Corvair. *The author*

engines used bushings instead of main bearings, which led to early burnout when the main bushings wore down and the aluminum blocks overheated. In 1961 and '62, the engine size of the Corvair was progressively increased, with a turbocharger added in 1962 to produce the ultimate sportster, the Monza Spyder.

Rear-mounted swing axle suspension created stability control problems for the Corvair. A suspension anti-roll bar was eliminated at a cost of $6 to help keep the base price low. Chevrolet required the over-wide front tires to be inflated only to a pressure of 15 to 19 lbs. to help correct for the swing axle oversteer problem. Tire overinflation by most owners caused the vehicle to become unstable. Advertising by competitors in 1960, which showed the results of shooting an arrow at the unstable rear end of the Corvair and missing its target by a large margin, caused a lack of confidence in many about the stability of the car. It was said "driving the Corvair could be a terrifying experience for the uninitiated — and was unpredictable even for the experienced." The original design of the car called for an anti-roll bar, but marketing and cost-cutting won out over intelligent engineering. GM never admitted to the stability problems.

Iacocca told me in my conversations with him that even though the use of the Italian name "Monza" on the sport model revved up sales of the car, he never felt threatened by it at Ford because the high-profile young lawyer Ralph Nader was making inquiries into the safety of the basic car and would eventually doom the car line with his book about its inherent problems. In 1965 Ralph Nader's book about the Corvair titled *Unsafe at Any Speed* exposed the stability problems with the car which could result in a rollover-type accident. General Motors' subsequent public attack on Nader's credibility apparently had little success.

Between the Nader allegations and with public confidence dwindling, Corvair sales began to drop seriously by 1965. The irony of Corvair's initial success as a sporty compact is that it helped precipitate a Ford car that would eventually do it in, named the Mustang. Chevy halted work on future Corvair designs due mainly to Ford's successful new Mustang entry in the small sporty car field.

Chrysler Corporation started concept design of its new small car in 1957 with a project named A901. The production version was to become known as the Falcon. However, just before the car was introduced in 1959, Chrysler found when they went to register the name in the Proprietary Name File of the Automobile Manufacturers Association, the official registry where all claimed automobile names are listed, Ford Motor Company had already claimed rights to the name Falcon. At the eleventh hour, the car was reportedly renamed the Valiant following the wishes of its designer Virgil Exner. He wanted it named after Prince Valiant from his favorite comic strip, *King Arthur*.

The Chrysler econo-box would be known simply as the 1960 Valiant and sold over 146,000 units. It was not advertised as produced by a separate Chrysler division, i.e., it was not a Plymouth, Dodge, or Chrysler. It was a stand-alone make. Part of the advertising theme that year emphasized the car was "Nobody's Kid Brother," and not a compact from any other Chrysler division. It was, however, sold through Chrysler-Plymouth and Plymouth-DeSoto dealerships.

1960 Valiant 4 Door Sedan by Chrysler Corporation. *Illustration by © Dan D. Palatnik*

In 1961 the Valiant was merged into the Plymouth line of cars and was renamed the Plymouth Valiant. The car was considered daring in its radical Virgil Exner styling and incorporated designs similar to the '57 Chrysler 300C grille and rear end from the Chrysler Imperial car. Early in the design stages, Chrysler stylists became aware of the new Chevy Corvair rear engine design. Rather than run scared and be intimidated by that new design, they settled on a new "Slant-6" engine up front where it should be, stretched the wheelbase so ample leg room was available, and stayed with Exner's unique styling. The new Chrysler 170-cubic-inch Slant-6 engine was unconventional in its own design. The efficient in-line six-cylinder engine was canted 30 degrees to one side with the transmission canted 30 degrees the opposite direction for balance. Its long branch individual intake and exhaust runners gave the engine superior performance and it quickly gained a reputation for durability and dependability. As a 1960 entry in the newly created NASCAR compact stock-car race at Daytona, Florida, the eight Valiants entered in the race placed 1st through 8th, beating European imports and V-8 models.

Valiants were built with unit-body, frameless chassis and had the first torsion bar front suspensions. In 1963 the car was totally reskinned and in the winter of 1964, just two weeks before introduction of the new Ford Mustang, a fastback-bodied Valiant was introduced as the Plymouth Barracuda, which many consider the first "pony car." Versions of the Valiant became the Dodge Lancer and Dodge Dart, which were introduced in 1963. But the new conservative boxy styling became boring compared to the soon-to-be-announced Ford entry, Mustang.

1960 Ford Falcon 2 Door Sedan. *Ford Motor Co.*

Ford introduced its new compact, known simply as the Falcon, on October 3, 1959. Falcon roots can be traced back to 1952, when Robert McNamara, then assistant general manager of the Ford Division, called for a research study to find what kind of people were buying the Volkswagen and why. As a result of the study, Ford management was convinced that a car sized between the Volkswagen and standard Ford could be successful. Henry Ford II committed the company to building this new so-called compact in 1957. McNamara had become Ford Motor Company's president by the time the Falcon was introduced and is considered by many to be the "Father of the Falcon." He left Ford Motor Company shortly after the Falcon's introduction, but his faith in the concept was vindicated with record sales of over 500,000 the first year and over 1,000,000 sold by the end of the second year. McNamara became Secretary of Defense in January of 1961, under the newly elected President John F. Kennedy.

The 1960 Falcon was first shown to the press on September 9, 1959. One day later, the highly publicized "Experience Run, USA" event was begun. This was an advertising program supported by Henry II to prove to the buying public that the new Falcon was a proven, tested and retested new car with unquestionable reliability. Of the first Falcons off the production line, 14 were sent on a trip with experienced endurance drivers covering every last mile of numbered federal highways in the country. The hundreds of thousands of miles racked up by these cars on the Experience Run and during tests gave Ford the bragging rights that the new Falcon was the "world's most experienced new car." The new lightweight six-cylinder overhead valve engine developed for the Falcon performed flawlessly. Test results from the run proved the 144-cubic-inch, 90 horsepower six could achieve up to 30 miles per gallon.

Body styles for 1960 included two- and four-door sedans, two- and four-door station wagons, and a Ford Ranchero car-based pickup transferred onto the Falcon platform. A direct derivative of the car, called the Mercury Comet, was also produced for the Mercury Division.

Later in 1962, the car was also offered as a Futura model with an upgraded bucket-seat interior and a changed roofline at the rear window to more of a Thunderbird design. These modifications were made to capture some of the Corvair mystique away from GM.

With the 1963 model, the new Sprint convertible model appeared later in the year with an optional new 260-cubic-inch V-8 with 164 horsepower and a four-speed transmission that was offered for the first time. The restyled '64 Falcon Sprint with a newer upgraded 289-cubic-inch V-8 and its brother, the similar sporty Mercury Comet S-22,

1962 Ford Falcon Futura two-door sedan with revised roof line. *Ford Motor Co.*

produced combined sales in 1963–64 of 191,000 cars. In those same two years, the Corvair Monza had sold 300,000 units. If Ford was going to catch up to Chevrolet, it would need a drastically new car redesigned from the ground up. Halfway through the 1964 model year, the Falcon was replaced by a sporty new four-seat car utilizing many of its own unique parts. That car was initially called a Special Falcon.

The cheapest Corvair available in 1960 had a factory price of $1,984. Ford, which had a long-practiced gambit of underpricing, priced the cheapest Falcon at precisely $10 less, at $1,974. *Car and Driver* magazine said, "Contrasting it with the Corvair, Ford has taken the opposite track from its major competitor and produced an absolutely normal compact car and in doing so has come up with something quite new. The Falcon was in fact the breath of fresh air the industry needed after the suffocating excesses of the late fifties." GM copied the Falcon chassis design by building the Chevy II, and later the Nova. Without those new platforms, there would have been no new Mustangs, Cougars, Camaros, and Firebirds. By celebrating simplicity, the Falcon set new standards.

These new compacts from the "Big Three" arrived in late 1959. First was the Corvair on October 2, then the Falcon on October 3, and the Valiant on October 29. The sales war commenced. Although sales of all three were brisk, it wasn't long before the Falcon became known as the "King of the Compacts," even though it was rather yawn-provoking. *Motor Life* magazine stated in December 1959, "The greatest accomplishments of the Corvair, Falcon, and Valiant are not listed in their specifications. It is the greater choice they have given the American car buyer. No longer is he restricted to one kind of a car, with variations in chrome or fins according to how much money he will pay."

After the initial introduction of the compact cars in 1960, Ford was busy contemplating new designs for the '60s. The "Big Three's" strategy of one-car-size-fits-all had been changed forever with the introduction of the '60s compacts.

The new decade was the age of youth, as 70 million children from the postwar baby boom became teenagers and young adults. The movement away from the conservative '50s continued and eventually resulted in revolutionary ways of thinking and real change in the cultural fabric of American life. In 1964 there were still "Whites Only" restrooms in Alabama. No longer content to be images in the generation ahead of them, young people wanted change. During the decade, women started wearing miniskirts, leather boots and fake eyelashes. Men wore paisley shirts, velvet trousers and long hair.

Americans had growing doubts about their automotive fantasies. They even developed a slang term for their ugly cars: "scuz buckets." The success of the increasingly popular Volkswagen made it stand out in the vast array of gaudy, gargantuan, look-alike sedans being produced by Detroit automakers. The sprouting countercultural youth generation had embraced the Volkswagen as the anti-car. It seemed to embody a denunciation of the homogeneity and the extravagance of the American car.

It was obvious now that the assumptions that had led stylists for over 30 years were seriously flawed for the given market. The answer in addressing the consumers' concern of auto homogenization was in giving the public a perceived differentiation of the same big family sedans by creating new auto types, using new structural and mechanical designs.

Car categories like the muscle car, the pony car and the high-end personal luxury car would soon appear. These would supplement the already in-production compacts Corvair, Valiant and Falcon.

The U.S. was launched into the Vietnam War at the end of 1960 in a struggle that would continue past the end of the decade. Significant world history would be made in 1961. Yuri Gagarin, the first man to orbit the earth, was shot into space by the Russians. The Berlin Wall was erected in Germany. The average annual salary in this country was $4,961.

By the end of 1960, Elvis Presley was home from two years in the Army in Germany, the Chrysler DeSoto was terminated and spy pilot Francis Gary Powers had been shot down in Russia. It was the same year John F. Kennedy beat Richard Nixon in the presidential election. A son named John F. Kennedy Jr. was born. Famous quarterback John Elway was born and Clark Gable died. Woolworth's began desegregating its lunch counters in the South. What a year it was, as history would show, and the compact car revolution was well under way in America.

And that's the way it was. These were the guiding factors Ford stylists, designers and managers had to deal with to satisfy the automotive needs of the lively ones, this new upcoming generation in this decade of the '60s.

Chapter 6

Ford Doldrums to the Formative Years

"People buy cars because they see in them something that turns them on."
Eugene Bordinat

On November 9, 1960, Chairman of the Board Henry Ford II promoted Robert McNamara to president of Ford Motor Company, the first person to be president from outside the Ford family. The same day, Lee Iacocca was promoted by Henry II to vice-president and general manager of the Ford Division, thereby filling McNamara's old position. Iacocca would now answer to McNamara. Lee Iacocca, the bright young engineer with a keen talent for marketing, had made his way to the top of Ford through the sales department.

Iacocca told me, "The reason Henry made McNamara president was because he was non-threatening to Henry." Five weeks later, McNamara would accept an offer to become the Secretary of Defense for President Kennedy. Lee Iacocca was immediately appointed by Henry II to fill McNamara's position as the new president of the Ford Division, the company crown jewel. Iacocca reported, "I only worked for McNamara for one day, and

Newly appointed Ford Division President Lee Iacocca, January 1960. *Walter P. Reuther Library, Wayne State University*

then he left. It was the shortest job I ever had!" But McNamara had taught Iacocca well over time. He told him, "If you have a good idea or concept, write it down — present it in written form — and it must be in 25 words or less — if you can't do it in 25 words, then you haven't thought it out well enough or you don't believe in it." These were words Iacocca would live by.

As Iacocca walked away from the fresh promotion in Mr. Ford's office, he found himself with thoughts "of that good looking little youth car" (as it was described in *Mustang* by Witzenburg) of which he had written in one of his pocket-sized black notebooks. This would be a car that could enliven Ford sales. Now he was in the right position and could do something about its creation.

Ford once again would begin a new decade absent a two-seat sporty car. The sexy '60s were about to begin and Ford had no cars to offer with true sex appeal. Iacocca said, "We at Ford began the '60s with a rather stodgy, non-youth image."

The late '50s and early '60s would turn out to be personally difficult years for the young Iacocca. A dedicated family man, his priority in life was family first and career second, with no deviations. His wife Mary developed insulin-dependent diabetes and, in the process of starting a family, had three miscarriages. Two daughters were born. First came Kathi in July of '59, who was immediately placed on the critical list due to a deep staph infection from which she made a quick recovery. Five years later, fragile baby Lia was born in July of '64 and had to remain in the hospital 10 days to gain enough weight to go home. (Lia, when asked in kindergarten what her father did, said, "I think he washes cars at Ford.") It was the comfort of family unity that would make the corporate pressures bearable during those career-topping years and the years to come. If there was a meeting to be missed for personal family issues, there was never even a decision to be made as to which received the priority.

The early '60s were the "modern" years, when a first class stamp cost a nickel. A family could have an entirely "modern" life, including modern houses, modern furniture, and modern dishes, but with one exception — the family car. Although the '50s mass-marketed cars were glitzy and meant to carry a family in formal style, they looked a lot more fun to drive than they actually were, with no help from those tailfins. Tailfins and aviation-inspired styling of the '50s were out. In the brief prosperity of the Kennedy years, buyers wanted styling, luxury and accessories. Those were the in things.

Brit Roy Lunn was working with Gale Halderman on Ford's first modern design of an electric car prototype. The vehicle was called the Firefly/Astrion and was conceived in the Special Projects Department. It was never produced. Technology was advancing fast. New boundaries were being discovered and Alan Shepard was launched into space as the first U.S. astronaut to ride atop a rocket.

A brief examination of the car parts after-market in the early '60s is appropriate because an entry by Ford had the potential of creating a huge sales marketplace. Throughout the first half of the 1900s, replacement vehicle parts were, by and large, an afterthought for most new vehicle manufacturers. Most replacement parts available to owners were produced by after-market parts companies. Ford Motor Company wanted to change all

Ford Astrion concept. *Ford Motor Co.*

that in the early '60s. Ford prevailed in an effort in 1961 to acquire the well-known and respected Electric Autolite Company, which was already producing numerous replacement parts for Ford vehicles. The parts company venture was successful and in 1972 the moniker was changed to Motorcraft, the name we commonly see today on Ford replacement parts. Ford Motor Company was in the original manufacturer replacement parts business for the long haul.

The Ford car division was offering an uninspiring mainline full-sized lineup plus the Falcon economy car. The elegant Thunderbird, sporting an entirely new design with cockpit interior styling by Johnny Najjar, was the only inspiring line. Iacocca surveyed the 1961 design offerings and found little on the horizon to meet the newly identified youth-oriented market. He felt Ford should define that market and lead the way by exploring design options to fulfill it. His aspirations for a specialty car for the youth market flowed over into Gene Bordinat's Advanced Styling Studio. The stylists' pencils were put to the task of creating two-seater and 2+2 sports car concepts.

As early as the summer of 1961, paper design mockups were beginning to line the wall of the styling studio. This was the year George Walker would retire as V.P. and GM of the Ford Design Center, relinquishing all his worldly design influence. Gene Bordinat's efforts had been well recognized and he was promoted to V.P. and GM of the Design Center, a position he would hold until his retirement. He utilized a strong management tool he had effectively developed. Every six months or so he would rotate stylists from studio to studio, and from interior design to exterior design, etc., just to keep them from getting "stale" in their ideas. Johnny Najjar agreed with that concept and gave credit to Bordinat for keeping their creative juices flowing.

Dr. Donald N. Frey.

The bright young engineer Don Frey, appointed head of Product Planning by Iacocca, realized that the generation of sex, drugs and rock 'n' roll was upon them — the Age of Aquarius was dawning. The *Northwestern University Alumni News* in 2002 quotes Frey: "I clearly remember sitting around the dining room table and my kids saying, 'Dad, your cars stink. They're terrible. There's no pizzazz.' And they were right." Frey decided that was going to change.

The '60s would become known as the last innocent generation. "I realized we were sitting on a powderkeg — or an opportunity," said Frey. It was time to ignite the baby boom market, the one that still prevails today. In January of 1961, Frey asked his Advanced Styling Department to draw up some plans for a little sports car. With the little two-seat clay model that resulted, Iacocca invited Grand Prix driver Dan Gurney and some other racing followers to view the model and give their opinions on the little car. Iacocca recalled: "All the buffs said, 'What a car! It'll be the greatest car ever built.' But when I looked at the guys saying it — the offbeat crowd, the real buffs — I said that's for sure not the car we want to build because it can't be a volume car. It's too far out" (quoted in *Mustang* by Nicky Wright).

He was referring to the 1962 two-seat Mustang Experimental Sports Car. We now start to see an "enhanced performance" image racing around in the minds of the stylists. With the 1961 model year well underway, little could be done to change those models across the board. Robert McNamara had left a rather lackluster image in Ford car building. As the lead time for planning and implementing changes for the new model year took several years, the '62 car models started out somewhat bland once again in the fall of 1961.

There was the return of the two- and four-door midsize Fairlane, the full-size Ford two- and four-door hardtops and sedans, and full-size Galaxies available with XL trim in the two-door and convertible models. Rounding out the lineup, there were also full-size Ford station wagons and once again, the Falcon two- and four-door sedans, station wagon and Ranchero models. V.P. Iacocca, due to long production lead times, was first able by the mid–'62 model year to have several sport-oriented changes incorporated into the already-in-production lineup. The two-door Falcon line received a new version called the Futura Sport Coupe which used a roofline similar to that of the Thunderbird, along with an upgraded interior. The midsize Fairlane gained a new version called the Sport Coupe, which had fancy interior trim, bucket seats, a console and simulated wire wheels. The full-sized Galaxie 500XL two-door and convertible models became available with bucket seats and a console with a floor shift.

To add a hint of pizzazz to the lineup, a big-block 390-cubic-inch V-8 with three 2-barrel Holley carburetors putting out 401 horsepower could be mounted to a Borg Warner T-10 manual four-speed gear box as an option in the full-sized Galaxie Starliner. This was planned just to have something to compete with Chevy's big new 409-cubic-inch Impala motor.

Jacque Passino, Ford's own internal racing promoter, had been pushing Ford brass to get back into high-performance car building. Through him, Ford teamed up with the Holman Moody Racing Team out of North Carolina. He also began the Indianapolis 500 custom race engine program during this period. Factory-backed racing was once again in swing. But that was about the extent of Ford's "performance lineup." These new production performance vehicles would be designed under the marketing catch phrase "Total Performance."

In a twist of fate, it was President Kennedy who set the wheels in motion for Ford's official return to racing. When Kennedy selected Ford president Robert McNamara to be his Secretary of Defense, it opened the door for McNamara's replacement, Lee Iacocca, who was willing to be daring. On June 11, 1962, Henry Ford II sent a letter to the American Manufacturers Association, withdrawing from a 1957 industry-wide agreement which banned factory-backed competition racing. That agreement by the early '60s was only loosely adhered to by the "Big Three," with all of them participating tongue-in-cheek on the weekends at informal races.

Ford's formal withdrawal from the agreement meant the "Total Performance" racing program at Ford Motor Company was off and running. Racing Director Jacque Passino furthered the company involvement with Holman Moody Racing. Within four years, cars powered by Ford would win the Daytona 500, Indianapolis 500 and the 24 Hours of LeMans. The doldrums at Ford were coming to an end, driven by the insatiable appetite of baby boomers for a new direction in U.S. car building.

On November 9, 1962, there was a tragic fire at the Ford Rotunda building, the mainstay for public relations, located at the headquarters of Ford Motor Company. The Rotunda, which was shaped like a stack of gears, was originally built as an exhibit building for the 1933 Chicago World's Fair. After the close of the fair, the building was disassembled, shipped to Dearborn, and reassembled on Rotunda Drive, across from the Central Office Building (the world headquarters at that time). The company would have various automotive exhibits in the Rotunda for the public and local school children. Tragically, in the course of roof repairs on that Friday in November, the hot tar roof ignited, and before the fire department could arrive, the building was destroyed.

Unfortunately, along with many other Ford records, the original production documents on the 1955 and some 1956 Thunderbirds were stored in the Rotunda, and all were lost in the fire. The Ford Rotunda in Dearborn was once the fifth leading tourist destination in the nation, ahead of such places as Yellowstone Park, the Washington Monument and the Statue of Liberty.

The Rotunda was not the only fire smoldering in the early 1960s. McNamara's influence had diminished at Ford, but was being felt elsewhere. As Secretary of Defense, he

1963 Ford Galaxie 500 convertible. *Ford Motor Co.*

approved the disastrous Bay of Pigs invasion of Cuba that failed to overthrow the government of Fidel Castro. Three years after the invasion, almost to the day in April 1964, the new Ford Division competition to his beloved Falcon, called Mustang, would be launched to the public and doom his compact car.

The '60s started out with no plastic bags, no air conditioning in most homes and cars, gas at 31 cents a gallon and Arnold Palmer dominating the game of golf. Chubby Checker started a dance craze called the Twist and the soon-to-be-famous *American Bandstand* dance show with Dick Clark began just before the invasion by the British rock group the Beatles. And another marketing sensation began in 1962; it was called Walmart. The '60s could hardly be labeled the doldrums.

Chapter 7

The Fairlane Committee

"There must be a market out there looking for a car."
Lee Iacocca

In 1960, even before his promotion to vice-president, Lee Iacocca had reasoned that if the flashiness and performance of costlier cars such as the Corvette and the Studebaker Avanti could be put into the shell of an inexpensive car for the masses, it would sell big time. "There must be a market out there looking for a car," he wrote in one of his little black books that had become his trademark. All he had to do was identify that market and build a car to fit it. The idea was hazy to him in late 1960 at the time of the promotion, and no actual footwork had been done to formalize the concept. A youth-oriented sporty car was the idea that raced about in his head. But was he on target? Was the youth market what he was looking for? Research was needed. He would need more formal input. He would need help.

Robert W. Hefty was in the Ford marketing research department as a public relations specialist. He had formulated two key demographic perceptions that would influence automobile marketing in the early '60s. A surge bubble of children born at the end of World War II as men returned home from war were coming of driving age in these years and would represent a new market. Their grouping would become known as World War II baby boomers. Secondly, with the prosperity of the era, many families were now able to afford multiple car ownership. Hefty's thoughts were becoming highlighted in Ford marketing circles.

In the early '60s, the Ford Division sales were 80 percent of Ford's North American auto business and profits. Iacocca was interested in improving the market penetration of Ford Division cars. Almost immediately after his promotion to vice-president and Ford Division general manager on November 9, 1960, Iacocca set up a hand-selected think tank. This was not an easy task, as people initially ran and hid in the closet when they heard someone was trying to develop a new concept car on the heels of the Edsel fiasco. Nonetheless, he and Don Frey picked a select group of Ford management types, which would represent market research, engineering, styling and product planning. For advice on advertising cues and implementation, a representative of the Ford Division advertising agency would also come to a meeting Iacocca had set to discuss a way forward with a new market segment he felt could be developed.

Lee Iacocca would head the committee. The son of Italian immigrants with a master's degree in mechanical engineering from Princeton would become known later by hobbyists as the "Father of the Mustang." He rejects the idea he was the "father" of the car and states it was certainly the result of a cumulative effort by all involved. "I surrounded myself with good men," he said to me.

His key man on the committee would be Product Planning Manager Don Frey. Frey was a married World War II Army officer with a PhD from the University of Michigan. A gifted, innovative engineer, he would become the project manager for the yet-to-be-conceived 1965 Mustang.

Hal Sperlich provided the committee keystone. At an early age of 31, he was a special projects manager for Don Frey. Even before the creation of the Fairlane Committee, Sperlich had conceived the idea of building a sporty four-seat car as a new addition to the Ford car lineup. My interview with him revealed his concept was guided by four guidelines he had developed: The car would need four seats; it had to be comfortable; it had to be good looking, with roadster properties; and it had to be inexpensive. The young Sperlich had no high-level management stature and would be unable to sell the idea without research guidance, but his thoughts became known to Iacocca. It was the Fairlane Committee that ultimately provided the research data to support Sperlich's idea of a new niche car. It was his idea to use the existing Falcon model platform and body parts for the new car. It's been said that without Hal Sperlich and this creative idea, the Mustang might never have been built with the limited budget eventually approved by Henry II.

Frank "Zimmie" Zimmerman, the pants-on-fire marketing guru, represented Ford Division Marketing.

Walter Murphy was a Ford public relations manager. He became responsible for creating the advertising program for the yet-to-be-created Mustang. He later became a personal confidant and PR man for Iacocca.

Sid Olson was from the J. Walter Thompson Advertising Agency, Ford's contract advertising firm. As a brilliant ad copywriter, he would work with Frank Zimmerman and John Bowers on the new vehicle advertising program.

These men completed the first tier of members for the committee.

Then there was Robert Eggert, Ford market research manager. With a PhD from the University of Minnesota, he would provide economic research for the validation of the project. He reportedly was the first to casually suggest the name "Mustang" for the car, a name which group research would also produce.

Chase Morsey, Ford car marketing manager, was the head of market research for the committee. The results of his research would have a profound effect on Ford marketing for years to come.

Jacque Passino, Ford racing director, provided high-performance and racing concepts for the committee to research. He kept pressure on Ford management to become reinvolved in factory-sponsored racing.

And John Bowers, the Ford advertising manager, was the Phi Beta Kappa member who would coordinate the new car advertising with Ford's ad agency, J. Walter Thompson.

The Dearborn Inn on Oakwood Boulevard in Dearborn was a long-established upscale hotel located just across the street from the current Ford Proving Grounds track and catered primarily to Ford employees, guests and visitors. It was built by Henry Ford in 1931 and was the first airport hotel in the United States. It was located on the site of the original Ford Airport, which was then located at what subsequently became the Proving Grounds and test track on the other side of Oakwood Boulevard.

Henry Ford had held many meetings at the Dearborn Inn, so it was an old stomping ground for Ford guys. Of course, it had an elegant barroom and drinks were served by white-uniformed waiters. But while the upper-echelon Ford "good ol' boys" were busy imbibing at the Dearborn Inn there was, several blocks away, a series of meetings going on that would become forever historically known in shaping the new Mustang.

A new synergy was developing within Ford that needed to be welded together to further this new niche concept. The ideas of Iacocca and Sperlich, together with Hefty's envisioned demographics, provided the spark to create this research group.

The first meeting of that committee was at the Fairlane Inn on Michigan Avenue. The informal name of the committee was derived from the name of their meeting place. This group convened to explore the new concept of an American-built two-seat sports car. The meetings appeared clandestine and were held in a private side conference room; Iacocca said they were not clandestine at all, but were just carried out in a convenient, uninterrupted way so as to get the participants together to form their strategies away from the turmoil of the office. Mr. Iacocca also told me he referred to the inn as the "Skunk Works" and that they needed to go off site so the "bean counters" weren't aware of what was going on. Some referred to these meetings as "secret" just so Mr. Ford wouldn't know about a new project that he would be so much against.

The Fairlane Inn on Michigan Ave., Dearborn, Michigan (circa 2007), **considerably changed in appearance from 1960s due to extensive remodeling.** *The author*

Newsweek reported later that as the project progressed, it "was turned over to a team of admen, engineers and designers working in the 'tomb'—a windowless room in an auditorium in the Ford Division office building where Iacocca imposed such rigid security that even the wastepaper was burned under supervision" (quoted in *Mustang* by Witzenburg).

Meetings initially were held twice a month and were considered to be future product planning meetings, or at least that was the way they were explained to those who inquired at the head shed. Eventually they became once-weekly meetings for an 18-week period. There was a lot of politicking at the meetings, all necessary to reach the compromises required to meet the committee goals.

One evening, committee member Walt Murphy was informed by the Fairlane Inn chef, one hour before the meeting, that the usual top prime steaks had not been delivered. In a panic, Murphy called the head chef in Henry Ford's executive dining room and asked if he could help. Murphy went to the 12th floor kitchen at the World Headquarters, threw 12 of the finest prime steaks money could buy into his leather attaché case, and quietly slipped out the door before anyone with the initials H.F. could notice. The group dined on the prime beef that night just as they did on the other nights, none the wiser.

As the group became more formalized in their meetings, the meeting time was changed from the evenings after work to 7:00 A.M. (sharp!) on Saturday mornings. The thrust of the group was twofold: (1) to allow Iacocca to present his idea for the need of a car for a newly envisioned market segment and to see if agreement could be reached on whether it was worth pursuing, and (2) for product planning to come up with a concept to fill the newly identified void in the Ford car lineup, if in fact there was one. All this was set against the backdrop of the Ford public relations department's being increasingly flooded with customer letters begging for the return of a car similar to the then-extinct 1955–57 Thunderbirds. Ford dealers' unhappiness with the mostly plain and uninspiring cars available in the early '60s was also a driving force. Amid Iacocca's cigar smoke and Walt Murphy's pipe smoke, the group would contemplate a concept that would become one of Ford's greatest successes.

Committee member Chase Morsey's carefully thought-out research showed baby boomers were just reaching car-buying age in the early '60s. He anticipated buyers between the ages of 18 and 34 would account for more than half of the projected new car sales for the upcoming decade. Additionally, car styling for the '60s would have to reflect the tastes of these new young buyers, not of the older generations. Young people had definite ideas about what they wanted in styling and performance. Iacocca said, "All the things youngsters want in a car are available on the market today—but not in the right combinations. They wanted the appeal of the Thunderbird, the sporty look of the Ferrari and the economy of the Volkswagen. But you can't buy a T-Bird for $2,500, get exceptional gas mileage in a Ferrari, or get whistled at in a VW. What they wanted, really, was a contradiction in terms."

Of all those under age 25, 36 percent liked the four-on-the-floor shifter feature and 35 percent wanted bucket seats. Research also indicated these buyers were becoming more educated, more sophisticated, and more willing to spend cash for what they called "image extension." Families had more money to spend with $10,000 annual incomes expected to increase 156 percent between 1960 and 1975. It was obvious more families would be able to afford second, third and even fourth cars.

Standing out in this group as the family members who most wanted their own cars were women and teens. From this research, it was agreed a potential and sizeable market did exist and it was made up of a young, affluent group big enough to create a substantial demand for a new kind of car, something distinctive and sporty with a dash of foreign flavor. It was also decided it should be small and maneuverable, it should have seats for four with room left over for a good-sized trunk, and it had to be priced for the mass market. That mass market consisted of four separate audiences: two-car families with cash surplus to spend; young drivers with very little money to spend; women who wanted a car that would be easy to maintain; and the sporty-set members who were seeking new fun toys.

This single new car was to be directed at these four previously separate market niches. They gave the project the code name T-5. The committee did not want to call the car a sports car, for it was to be more of a functional fun car, so the moniker became "sporty-car."

Definition of a future sporty car product to fill this newly defined void would be elusive, but from great minds come great ideas. The committee was actually created as a think tank to discuss direction in developing and creating a new theme car. Various departments would, as a result of these meetings, become involved in creative styling projects that would produce an almost prolific number of new concepts and designs.

There is a misconception among historians that needs to be cleared up. The Fairlane Committee was not established to create the Mustang. It was created to identify a new concept market. The Mustang car as we know it today was the end result of many different styling exercises, which included one that was named the Experimental Mustang Sports Car. The production Mustang introduced in the spring of 1964 as a 1965 model could be called the end result of this research committee's T-5 Project some three years earlier.

Although T-5 was the project name, the prototypes that would eventually lead to the 1965 Mustang had many names. These prototypes were never considered for actual production, but were used to help solidify concepts into a workable platform which would become the new Mustang. Different styling studies produced models named Median, Mina, and 12 different versions of the Avventura, the last of which became the Allegro, of which there were 13 versions. The favored Allegro design was modified to the extreme so many times it became stale to its designers. There was a Dave Ash original design concept called the Cougar that was eventually changed to another similar design called Cougar II. That was changed to Torino, and then to Mustang II.

But the one thing that became clear through all of this was that the drive train would be conventional — i.e., a front engine with a manual transmission with a driveshaft connected to a live rear axle. And the drive train of choice was already available and in production, installed in the six-cylinder Falcon car line.

Allegro 6/22/62 version. *Ford Motor Co.*

The informal Fairlane Committee ceased meetings at the end of 14 consecutive weeks with their two goals met. Iacocca was successful in identifying a new car buying segment and as a result of the committee's research, the impending seismic shift in the car market was revealed. Iacocca said, "It was the eighteen- to twenty-four-year-old group ... the accumulators, the career starters, the trend setters ... that we wanted to get to. We had to get into their minds even if they couldn't afford the car at first. Hell, we'd hit on such a good thing that we had to get moving on it before somebody else could come along and beat us to it." He had developed the research tools to fine-tune Henry's ear to the new four-seat sporty car concept that would create a revolution in Detroit auto building.

The greatest accomplishment of the committee was executive recognition of an era of youthful exuberance that existed in the country spread by its young President John F. Kennedy. This set the background for introduction of a new kind of car from Detroit. The members of the committee are recognized for their insight into discovery of a bottomless marketing arena and setting the stage for the Ford Motor Company introduction of a revolutionary new category of personal transportation for the young. They had little to do with the actual building of the new car called Mustang; they had everything to do with setting the stage for making its production possible.

The Fairlane Committee went down in automotive history as one of the most productive and innovative groups ever formed at Ford Motor Company. On September 10, 1962, the end result of the Project T-5 program was signed as "approved for production."

Chapter 8

Concept Sports Cars

> "Sports cars are by their nature controversial: they arouse the interest of the adolescent — and of those reaching second childhood; they excite the otherwise calm — and accentuate the egotistical; they are admired by many — and purchased by comparatively few."
>
> Roy C. Lunn, Project Engineer, 1962 Mustang

We need to understand just what a stylist is and how he or she thinks to see what leads to a unique design. Stylists have been described as artists, architects, sculptors and production designers. They have learned how to blend harmony, order and intent. They unite art with technology, and their capabilities are limited only by their own imaginations. The stylists of Ford design were unique in their own blending of interaction within the departments.

In the spring of 1961, Lee Iacocca asked styling chief Gene Bordinat to begin some styling exercises in his studios for two-seat and 2+2 seating configuration vehicles. As those concepts were coming together, Iacocca requested to have every Ford and Chevrolet model assembled in the styling courtyard for comparison. With all the cars lined up, model for model, facing each other, there was an obvious void just where Iacocca knew it would be — in front of the Corvair Monza.

Originally Iacocca wanted to build a 2000-lb. car for $2,000, he told me. But the engineers and stylists told him it was impossible to do. Pricing was very important to the entire concept. "For not much more than the price of a Volkswagen you could get so much more in our new Ford," he said of his theory.

The committee established a set of criteria for the new car they would eventually develop. It would have to be small, no longer than 180 inches; lightweight, no more than 2500 pounds; and inexpensive, not more than $2,500. They wanted a car that would carry four people. They wanted to retain the styling of the original, now classic Thunderbird, with a long hood, short rear deck and a low profile. It had to be offered with either a six- or eight-cylinder engine and it had to be versatile enough to be adapted to a wide variety of tastes.

Gene Bordinat already had several concept designs in the works deep in the confines of his styling department. These he developed with no program backing and they were unknown to outsiders. "We keep several of these exploratory type designs around in anticipation they might be useful in giving our management a new opportunity," said Bordinat in a 1985 Benson Ford Research Center interview.

Chevrolet Corvair Monza. *The author*

Allegro concept, August 16, 1962. *Ford Motor Co.*

Gale Halderman in the Advanced Concept studio worked with Hal Sperlich on different concepts they put together as hybrids of other concepts. These were known as "mule" cars. One of these was designated the Median Sports Car, which was a four-seat model. Continued work would evolve it into the Allegro X-car.

Another concept called Avventura was modified to become the Allegro. Bordinat said:

> In doing the car, the designers were looking for some type of styling clues that would give them a personalized sports car look. They had decided a long hood didn't look all that bad and it suggested horsepower. A four-seat car, it had a modestly compressed rear seat that was reasonably acceptable and the low overall height was good. Allegro was the first car in that general configuration. This concept was just what Iacocca had in the back of his mind. It was an interesting car and perhaps if it had been put into production, it could have done as well as the Mustang did, but we'll never know.

Iacocca wanted to see more variations on that concept theme, so stylists and modelers spent the next 9 months developing 13 different versions of the Allegro and other small sporty concepts that were two-seaters, 2+2s and four-seat cars that displayed varying amounts of rear seat legroom. One of the more sporty Bordinat designs was called the Cougar II. This prototype had decorative leather bucket seats with room behind for luggage, but no rear seats.

One of the first concepts the stylists came up with was a two-seater that would become later known as the 1962 Mustang Experimental Sports Car. Hal Sperlich, Special Projects Manager in the Ford Division's product planning group, said:

> John Kennedy was president and the country was taken by the enthusiasm of its youthful leader ... the excitement, the promise ... everything was upbeat and youthful at the time. Iacocca came in as head of the Ford Division, and he was a vibrant kind of guy, a go-go type who wanted to make his mark, and he seemed to fit all of that. It was one of those wild times when the chemistry of people was right, the times were right.

This two-seat concept car found fame and is widely known to Mustang hobbyists as the 1962 Mustang I.

Mr. Iacocca was already outside the box, thinking of other ways to come up with another car design. There was a clamoring by original two-seat Thunderbird owners to return the "little Bird" to production. He asked Tom Case, who was a product planner on the '55 Thunderbird project, to check out the possibility of creating a new four-seat car based on the design of the old two-seat Thunderbird. Don Frey directed the planners to contact the Budd Company, manufacturer of the original two-seat Thunderbird bodies. He learned Budd still had the original tooling for those cars. Frey's group presented their ideas to Budd management and jointly through Hal Sperlich they came up with a working prototype which used a cut-down '61 Falcon underbody with a modified '57 T-Bird body.

They named it the "XT-Bird." It could hold only two small children in the rear jump seat when the top was up. It looked more like a Falcon due to the removal of the characteristic tail fins and hood scoop. Even though it could be produced for substantially less than a totally new vehicle, it still looked like a Falcon and realistically only had seating for two adults. The proposed vehicle mockup was shown to Iacocca and he immediately

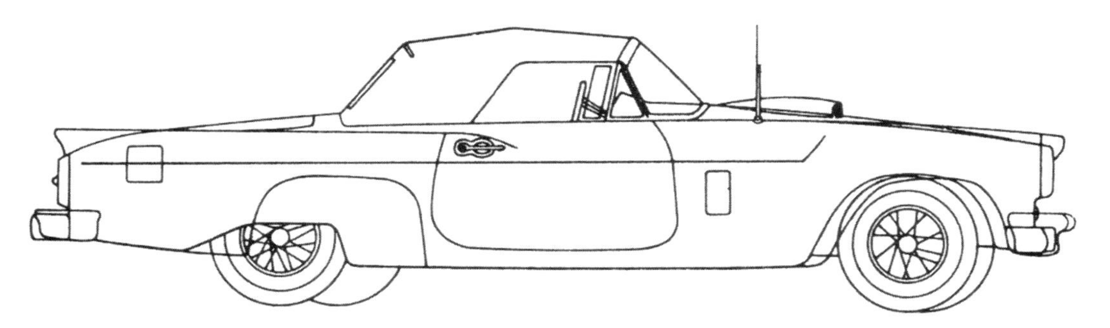

XT-Bird design overlay on 1957 Thunderbird body as envisioned by Budd Company. *Ford Motor Co.*

turned it down. He later said, "I was dead set against that car, because research had showed we needed a four-seat car in order to sell to the masses in the quantity we needed. I wanted a clean start." The one-of-a-kind prototype created for this exercise was later donated to the Henry Ford Museum by the Budd Company and resides there to date.

Through all these design concepts, one thing was painfully clear. The car had to have four seats with a realistically useable back seat for kids. The dimensions of the Allegro mockup were just about right for the kind of car they wanted. Early in '62 Hal Sperlich showed some of the designs of the concepts to Henry II at a planning meeting. "He was cold toward the whole idea and said he was not having any part in that kind of a program at all," said Sperlich. But, marching on with orders from Iacocca and Frey, Sperlich and Ford agreed that the seating package had been fairly well set. There was a lot of work ahead for the stylists trying to get a design that worked, with one failure after another. In the first seven months of the year, the styling department had turned out no fewer than eighteen different models.

Special Project Manager Sperlich knew the secret to successful economic cost projections for the car would have to include use of an existing platform on which to build. He had been quietly operating in the background from the time of the Fairlane Committee. His goal was to discover from existing production Ford cars a platform with potential for use on the new concept. He narrowed his search down to the Falcon car chassis after spending three full days surveying its traits. He figured a way to improve the floor pan and lengthen the car by adding extensions to the front end. This would allow for the roadster qualities they were looking for. The existing cowl assembly and windshield could also be included. That architecture would become the secret to unlocking the basis of the new car design.

The Competition

Frustrated by lack of good workable styling, Iacocca decided to call for a competition within the styling departments. It was now midsummer 1962, and the favorite project car Iacocca was formulating was not yet in an acceptable and economic design. Managers Iacocca and Frey decided unanimously to trash all the prototypes and begin with a new fresh design. There evolved a marketing brainstorm that would probably become the key to the whole concept of what would become the Mustang: a long, long option list that would allow the buyer to tailor his car for economy, luxury, performance, or any of these combinations.

On July 27, Gene Bordinat was given an ultimatum by Mr. Iacocca to produce six new clay models within two weeks, one of which would be chosen as a final working model platform. Hal Sperlich was chosen to write the rules for the design teams to follow in the competition. Bordinat said, "I turned all the troops in this place loose to come up with these models. We had our three studios within the Design Center work on these cars." The studios were Corporate Projects (Advanced), the Ford Studio and the Lincoln-Mercury Studio, all with a combined seven groups of designers. Each group was given the design dimensions established by the Fairlane Committee and were asked to engage in open competition.

Ford history shows the "challenge sent a wave of enthusiasm through the studios.

Avventura concept. *Ford Motor Co.*

Designers who normally shared ideas were locking their studios, refusing to let their rivals in. They all wanted to win that competition." In an amazing performance, within two weeks, by August 16, the studios turned in seven completed design models. Master design sculptor Bill Harbowy, who worked on the winning clay model, remembered in our interview:

> Iacocca would come out to the courtyard at the styling center with his entourage in their business suits and ties and circle the clay models with eagle eyes. Iacocca would remove his coat due to the heat and the rest would simultaneously also remove their coats. He would light up a cigar and some of them would light up cigars. The Ford Studio–designed "Stiletto" concept caught Iacocca's eye. He said, "This sleek little car was a beauty and we really wanted to go with it, but when we priced it out, we found we just couldn't build it for our price target of $2,500."

The other Ford Studio–designed car, known as the Cougar, was selected as the winner of the competition. Of all the other cars produced from that competition, none had the leanness of the Cougar. "They didn't have youth written all over them and they were a little too heavy—fat," said Iacocca. The Cougar was designed by stylist assistant Gale Halderman working under Dave Ash for Ford Studio head Joe Oros. The design

Joseph Oros. *Courtesy Mustang Club of America*

was done on a crash basis, with work on weekends, including Sundays, and a few all-nighters. As Oros explained it, "To save time, we put a list of dos and don'ts tacked up on the wall before we started sketching. We didn't want the car to look like any other car. It was supposed to be unique." Although Gale Halderman and Dave Ash are the ones who came up with the favored design, Oros is the one who usually gets the credit for it.

Styling chief Bordinat said, "The Cougar was the one that jumped out of the group. Some were sheer and some were soft, but this one seemed to have the greatest distinction. It was the one selected by management and the one we went on to produce." And it went from clay model to assembly line with very little alteration. Through its final design stages, it would become known as the Torino and, confusingly, the Mustang II, even though the prototype two-seat Mustang had not yet been completed. Although Oros went on to create a two-seat fastback version called the Cougar II first shown in late 1963, it was never seriously considered by Iacocca because it only had two seats and didn't meet the base criteria.

That was the last of the pre–Mustang concept car designs. The 1962 two-seat Mustang prototype was just becoming a completed reality that summer of '62, but as a completely separate concept project. That prototype had no influence either in the styling or in the way the final production 1965 Mustang would be built.

Chapter 9

Ford's Mustang Experimental Sports Car

The technically correct name for the concept design that eventually became known as the Mustang I prototype is the "Ford Mustang Experimental Sports Car." A detailed discussion about this 1962 styling concept is important, since there is confusion about its beginnings and how it fit into the overall picture of the development of the 1965 production Mustang.

This prototype had very little to do with the four-seat production version of the car that was introduced in 1964 as a 1965 model with a second-generation name. The name "Mustang" assigned to this concept car was not carried forward to the production car! The vehicle was built simply to test the performance-oriented market and explore potential excitement for a Ford-built sports car. It did help identify that market enthusiasm for Ford, and this important accomplishment is why we should remember this vehicle in Mustang heritage.

It might as well have been called the Eagle, or the Coyote, for example. The vehicle's name, Mustang, was — just by coincidence — also selected later for the production version. The production four-seat 1965 Mustang was *not* named after this prototype; it was just coincidentally of the same name. This is explained in detail in a later chapter.

In a slightly different version, the horse emblem hardware from the two-seat prototype Mustang is one item that indirectly transferred as a styling cue to the production car. Additionally, the side C-shaped air scoop design was incorporated into the production car styling.

Based on facts, it's best not even to try to make any connection between these two vehicles beyond the research influence toward market reaction for a two-seat sports car from Ford. The two-seat 1962 Mustang and the four-seat 1965 Mustang were two entirely *separate and distinct* programs, not related in any way. This is confirmed in my 2007–8 interviews with Hal Sperlich, 1965 Mustang project manager; John Najjar, 1962 2-seat Mustang executive stylist; and Gale Halderman, 1965 Mustang designer.

This unique concept had its beginnings in 1960 with Ford engineers Frank Theyleg and chassis engineer Royston Lunn. They discussed possibilities for usage of the German-built Ford Cardinal economy car drive train in a lightweight two-seat sports car. The Cardinal used a small front-mounted V-4 engine with a transaxle. Those early thoughts are about as far as that concept got until late 1961, when Lee Iacocca and his product planning manager Don Frey interested Henry II in searching for a way out of Ford's conservative marketing image. They engaged 42-year-old styling chief Gene Bordinat to begin the design of a market viable sporty two-seat car to fit into the newly minted marketing scheme named "Total Performance." They wanted a car the size of the British-made MG or Triumph sports cars.

In January 1962, Bordinat assigned two chief stylists in the Advance Studio, Bob Maguire and Damon Woods, to a project to develop a small sporty two- or four-seat car. Working for these two men were senior stylists John Najjar and Jim Sipple, along with a few other modelers and engineers. In March a new designer at Ford named Phil Clark joined the team,

Clark's styling influence in the actual chassis design may have been derived from a previous sketch he made from years of perfecting his own two-seat sports car designs. He also brought the design of a running horse emblem. This was a symbolic graphic he had developed over his career that would later contribute to the concept car design.

These are the men who created the clean-design, quick little two-seat sports car that would soon steal the hearts of all those who saw it. It was created in Johnny Najjar's small space in the Advance Studio, where many designs for all of the corporate divisions originated. With just a drawing on a blackboard, they gained approval of Gene Bordinat to build a full-size clay model and then a fiberglass prototype under the project name W-301. The program was then assigned specifically to the Ford Division as a Ford project. The model went from the sketch to final clay model in just 21 days by late May.

Herb Misch, the 44-year-old V.P. of Corporate Engineering, was looking for a show-stopping concept vehicle to be shown to the press that fall, at the 1963 model year introduction car show that would embrace a performance theme. Misch liked this concept model and decided to use it for the show. He selected a special projects manager named Roy Lunn to head the project and act as liaison between styling and engineering and to oversee the build of the vehicle.

The original plans which were produced in May show provisions for the installation of the Cardinal engine in the car, so, if the decision was made, it later could be converted to a drivable vehicle. And it would be. By the end of May 1962, final styling was cast in fiberglass. A formal program was initiated to build a drivable vehicle named the Mustang Experimental Sports Car. Ford now had the fiberglass prototype of a two-seat sports car,

Early Phil Clark sports car design. © *Phil Clark Archives*

Top—Actual Clark running horse emblem. *©Phil Clark Archives*

Middle—Design plan for the two-Seat Mustang. *Ford Motor Co.*

Bottom—Modelers styling the final clay design in the Advance Studio. *Robert Negstad Archives*

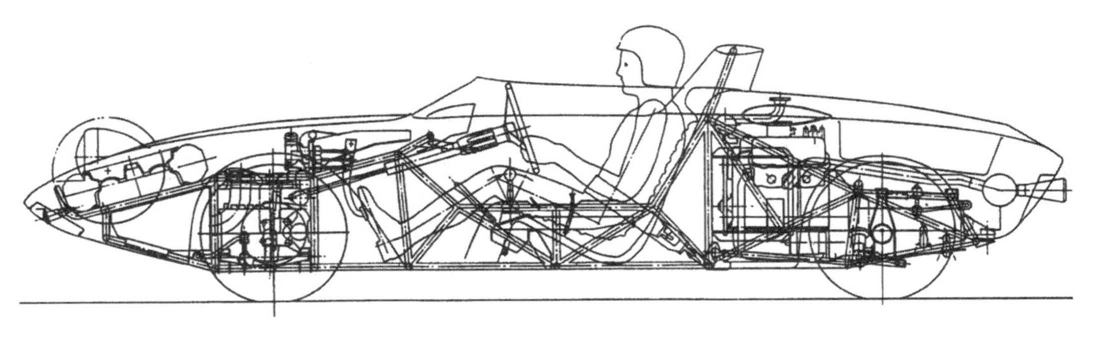

developed on participant enthusiasm, from a no-budget project. Bordinat and Misch decided they wanted to display the car at the U.S. Grand Prix race at Watkins Glen, New York, to be held October 7, 1962. Iacocca approved.

General Motors was going to build a new model of the Corvair Monza and it was to be displayed at the race, according to a tip received at Ford. It would have restyled bucket seats, a console and a fancy floor shift. The Ford boys decided they wanted to have something at the track to upstage the Monza, and they proceeded with their plans.

The time was short. Brit Roy Lunn, the ex-designer for Aston Martin, had quickly assembled a team to build a drivable car, including designer James Sipple and a bright young 32-year-old suspension engineer named Bob Negstad. He brought the concept of front-wheel drive to Ford and played a significant role in developing the front-wheel drive German-made Ford Taunus, an economical family car introduced in 1962 in Europe. This young group of engineers decided on a tube-type frame in combination with the 1500cc V-4 front-wheel drive Taunus engine/drive train uprated to 109 horsepower @ 6500 revs.

Two engine placement concepts were explored. A conventional front engine installation would allow shorter overall length and easier accommodation of luggage. Although the Taunus unit was for a front-wheel drive vehicle, it could also be used for rear-wheel drive, which would mean the engine could be installed amidship. Such an arrangement would allow a cleaner, low-drag front end and offered better weight distribution. The midships engine and rear transaxle placement easily adapted to the independent rear suspension, and was chosen for the prototype.

May 25 showing of the first completed styling example. *Ford Motor Co.*

Mustang interior design. *Ford Motor Co.*

As the suspension was developed by Negstad, it was decided to go all-out for performance, since the car would be driven around the track at Watkins Glen. Performance would be the name of the game.

The team was given a short 100-day time frame in which to complete a drivable car to be shown at Watkins Glen on October 7. The one-hundred-day clock began from the time they had a full set of blueprints at the end of June to the Watkins Glen show date. Ford designers had never used a stressed aluminum skin over a tube frame before and there was too little time to develop the concept. The search began to find help in fabricating the thin-skinned, tubular steel body, and they settled on a firm named Troutman & Barnes in Venice, California.

This custom race car fabrication shop, owned and run by Dick Troutman and Tom Barnes and popular in the '50s and '60s, was well known for an Indy 500 race car built for Rodger Ward. The shop also produced Scarab race bodies built for Lance Reventlow and Chuck Daigh, and the Chaparral race cars built for Jim Hall in the early '60s. The building they used to build these cars was the same building Carroll Shelby moved into later to build the first Shelby cars.

The "plug" fiberglass body mold was shipped to Troutman & Barnes on June 9, and with the help of craftsmen at California Metal Shaping Company, the aluminum body was created by forming it over the fiberglass body mold. By the end of June, Negstad and Lunn carried the final frame plans out to California, where Troutman & Barnes would build the tube frame chassis from the plans. The in-house manufactured frame was mounted on sawhorses, where workers skillfully attached the completed aluminum body in record time.

Designed into the prototype by Ford engineer Ray Smith were several unique features, like retractable front license plate frame and headlamps. He also designed fixed-position seats that relied on suspended accelerator, clutch and brake pedals hung from a stamped member that slid fore and aft four inches for driver-adjusted pedal comfort. Mustang was the first Ford design to use computer-aided design technology. The program was used to produce a suspension design protocol written in the new FORTRAN language

By mid–August, with barely a budget, a drivable version derived from the original fiberglass prototype was proceeding on schedule. The assembly was shipped back to Dearborn in late September for completion, after which it would be the centerpiece for the new "Total Performance" marketing plan.

The completed chassis arrived back at Ford on September 27. The project deadline of 100 days was met and the completed version was shown on October 2 at the Styling Department. At 154 inches long and weighing in at 1544 lbs., the car was dimensionally competitive with the MGs and Triumphs of the era.

Loading Mustang for flight to Watkins Glen Raceway in New York. *Ford Motor Co.*

1962 Mustang at Watkins Glen Raceway driven by Ford suspension designer and engineer Bob Negstad. ***Bottom***—Dan Gurney in the pit on the Watkins Glen track. *Robert Negstad Archives*

The car was flown that same day in a Ford company airplane to Elmira, New York. There it was loaded on a pre-positioned Ford tractor-trailer truck that had the Mustang name and horse design emblazoned on its sides, and then was driven to Watkins Glen Raceway.

Negstad said, "We took the car to Watkins Glen, where famous race driver Dan Gurney drove it for a couple of laps where it recorded speeds over 100 mph and acceleration of 0–60 mph in 10 seconds." Thousands of spectators were impressed on race day.

People saw the car and said, "YES!" In contrast to their reaction to the roped-off Corvair Monza display car, which was not even drivable, the crowds continuously surrounded the Mustang with no real attention given the Monza. Victory for Mustang! There was never any real intention of putting the Mustang into production for many corporate reasons, according to Iacocca. But he had found what he was looking for with this little car — market enthusiasm — which Fairlane Committee research told him was there. For this little experimental prototype, mission impossible was admirably accomplished.

Iacocca said, "This first Mustang was rigged up to go to Watkins Glen to show the

Rare view of Mustang with headlamps extended at Watkins Glen. *Ford Motor Co.*

kids that they should wait for us because we had some good hot stuff coming." Little did they know that Ford was rushing a hot new four-seater into production simultaneously.

The 1965 four-seat Mustang had already been designed and its production approved one month before the Watkins Glen showing of the little two-seater. The four-seat production version styling was finalized on September 10 and the two-seater was shown to the public first on October 5, further evidence that there was no styling influence carried over to the production car.

Finding the Mustang Name

A common misconception is that Johnny Najjar named the car after the World War II P-51 fighter plane named "Mustang," manufactured by North American Aviation in the 1940s. In my March 3, 2008, interview with Mr. Najjar, he stated:

> R.H. Bob Maguire, my boss, and I were looking through a list of names for the car. I had been reading about the P-51 Mustang airplane and suggested the name Mustang in remembrance of the P-51, but Bob thought the name as associated with the airplane was too 'airplaney' and rejected that idea. I again suggested the same name Mustang, but this time with a horse association because it seemed more romantic. He agreed and we together selected that name right on the spot, and that's how it got its name.

Bob Maguire, a Mensa Society member, came from General Motors as an interior stylist in the early '50s. It was in his office that John Najjar, as a joke that backfired, set off an exploding rocket in the early '50s. Maguire, who had been hired by Bordinat, was heavily involved in the design of the 1955 two-seat Thunderbird and had become a department head under Gene Bordinat. Although he was known for having no sense of humor and was an accomplished classical musician, he was remembered more by many as the guy who took a styling prototype out on the road with no license plates and was charged with driving with no license plates, speeding and failing to stop for the police. He spent a night in jail for that. Among his many good accomplishments at Ford, he is credited, along with Johnny Najjar, with assigning the name to the two-seat Mustang.

It has been recently reported by Holly Clark, author, daughter of Phillip Clark and custodian of the Phillip Clark Archives, that according to Clark's records, he brought the name Mustang with him to Ford from his previous seasoned drawings of horses, a name he used on the drawings even as far back as when he attended the Art Center School in California in 1958. His suggestion of that name may have played a part in Maguire's thoughts about name selection.

The official Ford press release dated October 7, 1962, states "the Mustang is aptly named: Mustang horses are small, hardy, and half-wild. The diminutive two-seater just trotted out of the Ford Motor Company stable fits the description."

A point to remember: *none* of the Mustang prototype cars or the 1965 production version was *ever* named after the P-51 fighter plane! Lee Iacocca confirmed this with me in a recorded interview in the summer of 2006. Had Johnny Najjar never *mentioned* the idea of naming the prototype after the airplane, along with many other suggested names also never selected, the correlation would have never been made.

The prototype is now retroactively and commonly referred to as the Mustang I since other vehicles were subsequently also given the name Mustang. The concept name technically became "Ford's Experimental Sports Car, The Mustang," as referred to in early Ford specification literature.

Designing the Mustang Emblem

There was no name assigned to the little prototype early in its development. It started as Project W-301, and then evolved into Mustang. In late May it was referred to as the "Mustang" and the "August Sports Car," "August Sport Car" and the "August Show Car." At Troutman & Barnes it was referred to and built under the names "August Sports Car Program" and "Mustang Sports Car Program."

Bob Negstad, the driveline and suspension designer for the car, related directly to me in an interview with him on April 25, 1999, how the emblem design came about. He was there that day. "While surveying the development of the car design in the styling studio one day, the design team began looking at a wall in the studio that was covered with drawings of concept emblems available for use by the studio on prototypes. Johnny Najjar, the project manager, stated he thought the car should wear an emblem that focused one's thoughts toward (1) a horse, and (2) a car that was American built. Use of the chosen horse-denoted name conformed to current usage by Ford of animal names on other cars, such as 'Thunderbird' and 'Falcon.' To convey the American-built theme, they should look toward an emblem that suggested patriotism."

Preliminary Mustang emblem concept by Phil Clark. © *Phil Clark Archives*

On the wall was an image, among all the others, of a horse with a vertical bar running through it. Attention was drawn to this particular image because it had a horse, half of the equation for the image for which they were looking. Stylist Phillip Clark removed it from the wall for further comments. When it was decided the horse didn't convey the desired motion, he retraced it this time with more of a "galloping" form and used more of a wind-blown long flowing tail trailing the horse. The first 3-dimensional emblem was made by Clark from cutout cardboard, wrapped in aluminum foil, and displayed on the side of the clay car mockup with an uncolored vertical bar.

Najjar agreed the horse now conveyed more action, but he asked, "How do we know it's an Amer-

First cardboard and foil emblem as installed on the clay model. *Robert Negstad Archives*

ican horse, and not an Austrian or South American horse?" From Clark's archived notes we know he had produced ten other versions of the design before settling on the design being considered. Clark rethought this design, and all of his previous thoughts about the design, and noted the non-distinct vertical bar running through it. He added several lines to create the appearance of three joined vertical bars and added color to each of the three bars: red, white and blue. "There," he said, "now it's an American horse."

Agreement came quickly that the image reflected both specifications for the design and the image was accepted as the one from which the emblems were made for the Mustang prototype. Clark, now long deceased, sketched horses as a favorite subject in his drawings, and his sketches usually showed the left-side view of the horse.

This red, white and blue emblem was used only on these two original two-seat Mustang prototypes and then the next year was used on the one-of-a-kind 1963 Mustang II four-seat prototype. The hard-cast prototype emblems used were actually made of two pieces: the pony, and the backing vertical-bar plate, to which the pony was then attached by screws.

The emblem was modified to a running horse rather than a galloping horse design and as a one-piece casting for the production 1965 and later Mustang cars.

The May 29th full-scale blueprint titled "Side View Check Chassis Layout," relating to the "August Sports Car," depicts a full-size driver positioned in the seat. He wears a racing helmet with a Thunderbird engine style emblem symbolic of the times, and with the numerals "90" in the center section, as can be seen in the drawing below. Ford was using

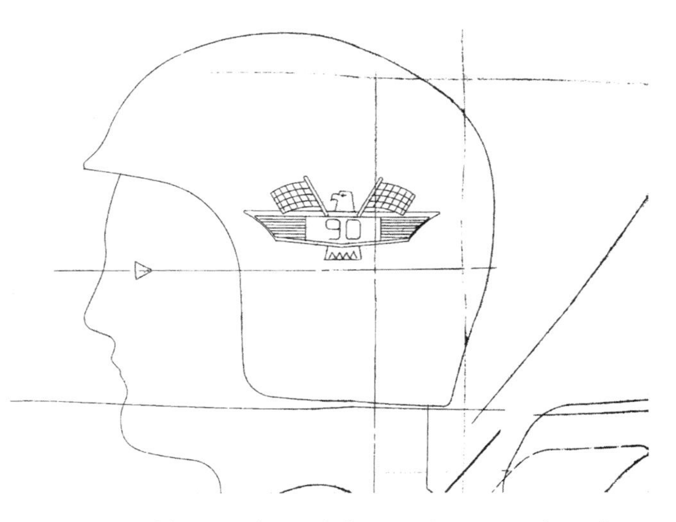

Prototype emblem on driver helmet as it appeared on the Mustang full-scale blueprint. *The author*

a similar emblem on its full-size cars in 1962 that numerically displayed the engine size of the car, e.g., "390" meaning 390-cubic-inch engine displacement. If the "90" on this helmet depiction is converted from cubic inches to cubic centimeters, the number 1475 is derived, presumably denoting the 1500 advertised cubic centimeter value of the Cardinal engine to be installed in the prototype. When these blueprints were made, one school of thought may have been to use this emblem denoting engine size on the side of the Mustang, as it was used on other Ford cars, in addition to or in lieu of the running horse emblem. Stylist Phil Clark may have had a hand in designing this emblem, or it may have been just from the imagination of a design engineer laid out on paper. It could have become the emblem used instead of the familiar horse. We'll never know, as it was never used as an emblem on the car.

Prototype Exhibits

After Watkins Glen, the car was shown at Laguna Seca Raceway in Monterey, California, in late October 1962. It again wowed the crowds with its speed and handling agility as demonstrated once more by driver Dan Gurney.

Shipped cross-country, it was driven at the Daytona Motor Speedway by Bob Negstad for the pleasure of NASCAR president Bill France to observe. The Mustang was displayed at auto shows that winter in Milwaukee, Chicago, Denver, Philadelphia, Los Angeles and Dallas, where it was seen by an estimated 1.5 million people.

One trip to California called for the car to be flown as air freight. Ford insisted an employee accompany the car en route, but air regulations prohibited passengers traveling with the cargo on any cargo aircraft. After some creative research, it was determined that people were allowed to accompany animals traveling as cargo on aircraft, so a parakeet and cage were purchased and placed in the front seat, and the Mustang was flown to California accompanied by Ford mechanic James Augenstein.

At the invitation of college officials at Purdue University in Indiana, the car was displayed for engineering students on February 12, 1963. The Ford "Report on the Mustang's College Tour" stated "the student turnout was so enthusiastic (with 400 students attend-

Mustang transporter arrival at Watkins Glen Raceway, New York. *Robert Negstad Archives.* ***Bottom***—The only two 1962 Mustangs ever produced (one a fiberglass model) with VP Herb Misch seated and VP Styling Gene Bordinat standing. *Ford Motor Co.*

ing) that plans were started immediately to bring the Mustang to college campuses throughout the country."

The primary audience targeted during the tour was engineering students at key universities who might be favorably influenced toward a career at Ford Motor Company. The Mustang was transported by road around the country in its special trailer pulled by a 1963 Ford station wagon and displayed on campus in a roped-in enclosure. Between April 3 and May 31, the Mustang was displayed at 17 different college campuses, including Penn State, Ohio State, Stanford, University of California Berkeley, UCLA, University of Southern California, Northwestern, and Michigan State. The report shows the car was seen by 63,300 students. The *Los Angeles Times* reported on May 5 that even though Ford said the car would never go into production, Ford had as of that date received 500 deposits to buy a production version of the car.

After the college tour, Ford wanted the little car to have exposure in Europe to study the reaction of the European public to Ford's version of the two-seat sports car concept, a concept Europeans had originated. It was flown to Europe for that purpose. "Mustang spent an entire year touring Europe, and the European press lavished continuous praise on it. We thought we'd never see it again," said Negstad. It was shown in the cities of Copenhagen in Denmark, Oslo in Norway, and Geneva in Switzerland.

After the European tour, the car was flown back to Dearborn, where it was received in poor shape from abusive handling and driving. Damage included dents, dings, chipped and scratched paint and various mechanical problems.

End of the Road

The little derelict became a vestige of an earlier design concept. Ford had moved on to the GT40 racing program and their other production performance cars. There was only one road left for the little Mustang, and that was the way of all other preceding Ford styling and prototype designs: to the smasher. Those stylists and designers who had created the Mustang couldn't bear to watch its crushing demise, so they hid the little car in a nondescript trailer behind a building on the Ford property. Every so often, some management type would ask, "What's in that trailer out there?"—and one of the guardian engineers would carefully hitch it up and move it somewhere else out of view.

Eventually, its existence was brought to the attention of the Henry Ford Museum and it was relocated to the museum in Dearborn. On November 3, 1975, it was officially donated by Ford Motor Company to the Edison Institute collection for permanent display at the museum. In 1980, for the retirement party for Gene Bordinat, the designing engineers took it out of its mothball storage, restored it, and made it the centerpiece display for the party. After the party, Ford Motor Company transferred it back to the Museum, where it remains on display to this date.

It is widely believed, but unsubstantiated, that two drivable versions of the car were actually built. Mustang Project executive engineer in charge Roy Lunn stated in our April

14, 2008, interview that the original car was only a fiberglass mold of the clay model that was finished to look like the final product. It was used as a static display model at shows, which freed up the drivable car for road trips and college campus demonstrations. It was never an operable car since it had no drive train or viable suspension. There were photos of the two cars taken together, seeming to suggest there were actually two drivable cars, but in fact one was really just the fiberglass model. First photos of that fiberglass model were taken on May 25, 1962.

The whereabouts or disposition of that fiberglass model is unknown; however, a *Car & Driver* reporter stated to me in April 2003, "The roll-around fiberglass model is stashed away somewhere in a Ford warehouse." That is pure conjecture at this time, as there are no records of its existence. Nor have any public records of its destruction been located.

The two major points to be remembered about the two-seat 1962 Mustang Experimental Sports Car are these:

(1) The vehicle was one of many concepts developed from Fairlane Committee research that led to numerous designs and prototypes incorporating ideas to be used as stepping stones, all leading toward the production of a four-seat car, which eventually in 1964 was produced and became known as Mustang.
(2) This two-seater design contributed virtually nothing to the design of the '65 production Mustang. It simply was one of many prototypes studied and subsequently discarded.

Ford's first experimental sports car put out to pasture.

The *one* physical piece conceivably contributed to the production car was an early version of the three-bar Mustang horse emblem. Gale Halderman stated to me in October 2007 that he may have referred to the emblem when he designed the production version, but it had no influence on him as he came up with his own design for the production car.

The name Mustang would again be used one more time, on a prototype car built in 1963 that became known as the Mustang II. After the Mustang II, the name was retired from concept usage. For all practical purposes, that was the end of the Mustang prototype lineage relating to the later '65 production version.

The two-seat Mustang prototype should more correctly be remembered as the first step of a movement by Ford to enter the formal racing circuit. Its design led directly to the development of the 1964 Ford Mark I GT40 F.I.A. GT Prototype. This is where the real heritage unfolds.

After Henry II's talks with Ferrari about buying that company fell through, Roy Lunn was made manager of the Ford Advanced Vehicles Group and was immediately assigned to design development of the GT40. "The lessons learned about computer-aided suspension design carried on into the big racing era cars and it all started with the little two-seat Mustang all those years ago," said suspension engineer Negstad. The mission for the little Mustang was completed. It did exactly what it was supposed to do — boost Ford's "Total Performance" image until the real production and racing cars could be brought to the scene.

The December 1962 issue of *Car & Driver* magazine proclaimed this Mustang "as the first true sports car to come out of Dearborn."

Chapter 10

The Performance Era

Ford, once again, resumed factory sponsorship for a NASCAR race team out of Charlotte, North Carolina, named Holman Moody Racing. This was a relationship that began in 1956 but ended formally with the 1957 Automobile Manufacturers Association (AMA) ban on factory-sponsored racing. None of the "Big Three" manufacturers loyally supported that ban after '57, but they quietly and unofficially were supporting race operations to boost sales of their own performance cars.

Ford restarted the sponsor relationship in late 1958, somewhat clandestinely, due to the efforts of Jacque Passino, who had kept pressure on management to re-enter the factory-sponsored racing scene. Ford gave Holman Moody ten stripped '59 Thunderbirds off the assembly line for modification to NASCAR racing specifications in support of their racing program. These cars were used with some notoriety for several years. By 1961, full-sized Fords modified by Holman Moody were used at tracks around the east-coast NASCAR racing circuit with disregard for the AMA racing ban.

One day in 1961 a young, smooth-talking race driver named Carroll Shelby walked into Lee Iacocca's office. With Don Frey in attendance, Mr. Shelby presented a pitch that if given a few engines, he could build a high-performance production sports car that would blow the pants off the Corvette. Prior to his approaching Ford, his idea was rejected by other car manufacturers. He got his engines from Iacocca and was proved right with his prediction. In retrospect, Iacocca jokes, "He talks so much, the only way I could get him to stop talking and out of the office was to give him the engines!" Carroll Shelby was now on board at Ford, using the newly developed lightweight Ford 260-cubic-inch V-8 in an English AC-bodied car he called the Cobra.

The first Shelby Cobra prototype roadster had its initial test run in England on January 30, 1962. By the end of 1963, the road-racing Cobras would win most amateur and professional events of that season. Importantly, it showed Ford was dead serious about reinvolvement in performance, and for the first time, it whetted the appetite of car enthusiasts to look for excitement coming from Ford.

Ford was moderately successful in the performance arena with the 1962 model year introduction of the 406-cubic-inch V-8 engine, bucket seats and consoles with floor shifts on four-speed transmissions in the big Ford Galaxies. A new stylish Thunderbird-inspired roof line appeared on the 1962 Falcon. Despite these and other styling changes, the younger set still looked at these Fords as "less than inspiring" visually.

1962 Shelby Cobra. *Courtesy of Lynn Park*

Henry II was President of the AMA in mid–June of 1962 when he formally made the announcement that Ford would no longer support the AMA ban on factory-sponsored racing since it seemed to "no longer have any purpose or effect." This merely was a formal way of saying that Ford would do exactly what GM and Chrysler were already doing in a not-so-secret way: participate in factory-sponsored NASCAR racing. Now the door was open for Ford to reenter racing in a big way.

In mid–1962, just to show that it could be done, and to add to the newly introduced sporty '62 Falcon aura, Holman Moody built a Ford-sponsored race-prepared Falcon called the Challenger which ran in the "12 Hours of Sebring" race and became known as the "Fastest Falcon on Earth." This was in addition to running 406-cubic-inch Fords at east-coast NASCAR tracks. By now, the Holman Moody name was one of the most highly recognized and respected names in NASCAR racing, winning race after race in Ford-powered cars with the word F-O-R-D spelled out in big letters on the fenders and hoods.

The model year 1963 would be an exciting one for Ford. The company marketing theme from '62, "The Lively Ones from Ford," would be replaced with the bolder theme "Total Performance." An in-charge Iacocca, who wanted involvement in performance programs, would refer to this as an all-out "crossed flags" campaign.

The '63 model line up was introduced in the fall of '62. The Galaxie 500 was available with the optional 406-cubic-inch Thunderbird High Performance engine with tri-power carburetion and a thumping 405 horsepower and mated to a four-speed manual trans-

1963½ Ford Falcon Futura Hardtop with the new sweeping roofline. *Ford Motor Co.*

mission. The top-of-the-line Galaxie 500XL had glorious cockpit styling with beautifully designed aluminum trimmed bucket seats, a floor-mounted shift on a grille design console, bright metal Mylar-trimmed door panels with carpeting, and bright metal trimmed brake, clutch and gas pedals.

As part of the fall introduction of the new '63 model line, excitement for performance was generated with the first public showing of the

Ford 406-cubic-inch engine/405 horsepower. *The author*

1963 Ford Galaxie 500XL Convertible. *The author*

1962 two-seat Mustang experimental sports car at Watkins Glen. Ford engineering Vice-President Herb Misch and his internal public relations man, Cog Briggs, were the ones who pushed the '62 Mustang prototype through to fruition for this sole purpose. The Iacocca "Performance Plan" was all coming together at the end of 1962.

Change for '63 became the Iacocca watchword. The earliest the full effect of Iacocca-influenced styling could be realized was early in 1963. A new concept in marketing by introducing mid-year models was put in place and these cars were marketed as 1963½ Fords.

Ford would launch these '63½ mid–year car models in January 1963, at an introduction time that was unprecedented for Ford, as well as the rest of the industry. This was unique marketing strategy by the Ford Division General Manager named Iacocca. Not only did he not have to wait until the fall to introduce these hot new models, he could launch them mid–year with all the automotive press focused only on Ford's new performance cars with little distraction from other manufacturers. Sales took off with the half-year new sloping roof lines and high-powered performance engines. In the spring of 1963, Ford offered an additional ground-pounding 427-cubic-inch Thunderbird High Performance engine with either a single four-barrel carburetor at 410 horsepower or with dual four-barrel carburetors putting out 425 horsepower.

Top left—**Lido "Lee" A. Iacocca.** *Lee Iacocca Archives.* ***Right***—**Harold "Hal" Sperlich.** *Courtesy of Harold Sperlich.* ***Bottom—(left to right)*** **Brothers Benson, Henry II, and William Ford.** *Ford Motor Co.*

 Top— Ford Motor Company World Headquarters, Dearborn, Michigan. ***Bottom***— Proposed 1962 Mustang prototype. *Ford Motor Co.*

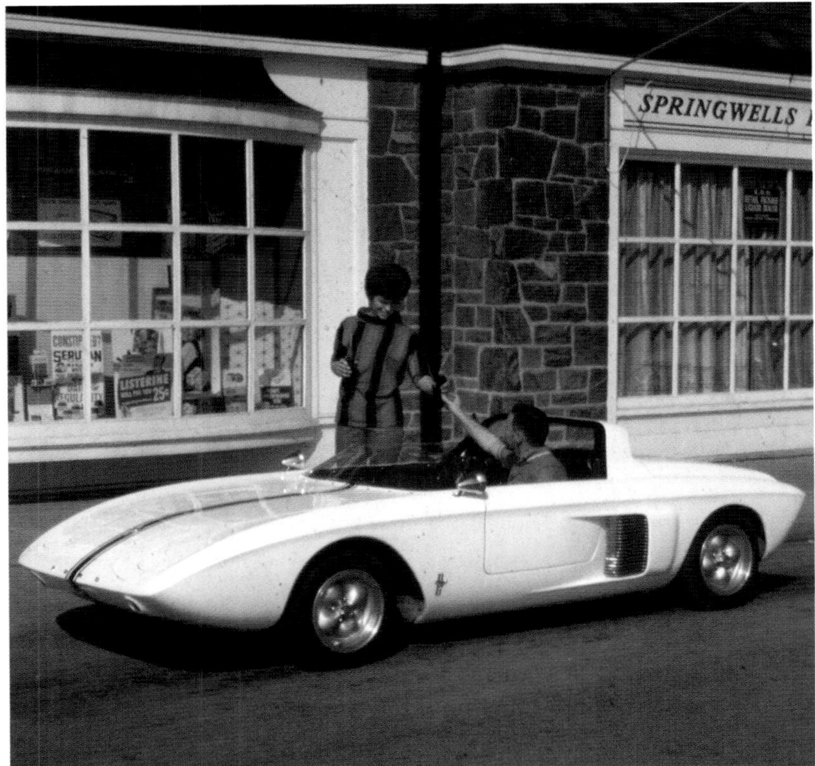

Above— Cougar II prototype. *Ford Motor Co.*

Right—The 1962 Mustang Experimental Sports Car. *The author*

Above—1962 Mustang arrival at Watkins Glen Raceway, New York, in the special transporter. *Robert Negstad Archives.*

A painting by Scott O. Kennedy, created for a 2007 Reach Foundation gala recognizing Lee Iacocca, portrays the impact of Iacocca and the Mustang on American automotive manufacturing. *© Scott O. Kennedy (www.scottkennedy.com).*

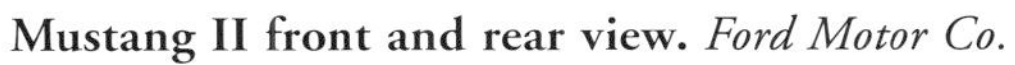
Mustang II front and rear view. *Ford Motor Co.*

Interior view of the 1963 Mustang II. Note the unique instrument panel and console. *The author.*

One of the first published new Ford Mustang advertisements simply highlighting the "Unexpected" and "$2368" low-cost themes. *Ford Motor Co.*

Top —New 1965 Mustang displayed on first floor Pavilion exhibit. © *Bill Cotter (worldsfair photos.com).* ***Bottom*** —1965 Wimbledon White Mustang Convertible typical of the ones used on the World's Fair Skyway ride. *Ford Motor Co.*

Top —First pre-production 1965 Mustang hardtop to be assigned a VIN, 5F07U100002 with its current owner, the author, retired United Airlines Captain Robert A. Fria. ***Bottom*** —Lee Iacocca with first pre-production 1965 Mustang hardtop 5F07U100002 (March, 2000). *The author.*

1963½ Ford Galaxie Fastback. © *Artwork of Danny Whitfield*

The newly added Galaxie model would become known as a '63½ Galaxie Fastback, which could also be optioned in XL trim. It was restyled with a sweeping, sloped roofline not only for good looks, but for better aerodynamics on the NASCAR race circuit. Iacocca-inspired styling and performance were now in full bloom. And as for the introduction of the new Ford 260 and 289 V-8's, Iacocca said, "It was the greatest engine we ever made. We put it in racing and we were winning everything we put it in."

Ford 427-cubic-inch engine/425 horsepower. *The author*

Top: **1963 Galaxie raced by Dan Gurney.** ***Bottom:*** **1963 Galaxie raced by Fred Lorenzen.** *Ford Motor Co.*

Ford would sell 1,462,100 of the 1963 model year cars before the year was over. But still missing from the '63 model year lineup for Ford Motor Company was that ever-elusive four-passenger sporty type car of which Iacocca had long dreamed.

With the new "Total Performance" Holman Moody race-modified 427 Fords, and '63 Mercury Marauders, wins were common all over the circuit with famous drivers like Dan Gurney, Fred Lorenzen and Fireball Roberts. The 427 Fords captured 1st through 5th place at the Daytona 500 race. A fastback 427 single carbureted car was clocked at 161 mph. These efforts by Holman Moody paid off well for Ford as their sponsor.

Henry II was now hooked on racing. In the spring of '63 he decided to purchase the Ferrari name, and with it, the whole factory in Italy. He formulated a plan for Ford to take over the factory to build the race cars and for Enzo Ferrari himself to campaign the cars. Henry wanted to own the plum of European racing so that nothing would stand in the way of world-class racing championships for Ford. Only problem was, Enzo didn't see it that way. A purchase plan was presented to Ferrari factory owner Enzo, to which he said "No deal," according to Fairlane Committee member Hal Sperlich, who participated in the Ford on-site presentation with Henry II.

Henry was infuriated. He decided if he couldn't have Ferrari, Ford would build their

Ford GT 40. *Ford Motor Co.*

own car to campaign against Ferrari and claim Ferrari world racing titles as their own while embarrassing Enzo Ferrari for his refusal to sell. Two-seat 1962 Mustang stylist Roy Lunn was immediately assigned to design a new race car made to win in Le Mans–style racing. The now-famous GT40 was born and relied on some of the engineering and styling that went into the original Mustang prototype. First raced at Le Mans in 1965, the GT40 would be continually improved, and in 1966 it was driven to 1st, 2nd and 3rd place at the Le Mans race in France, beating Ferrari. Henry had gotten even, as he saw it, with Enzo.

Ford race sponsoring was continued right into the 1964 season with 427-powered '64 Fords and numerous wins. With more winning drivers like Mark Donohue, Bobby Allison, Parnelli Jones, Ned Jarrett, Peter Revson, David Pearson, Al Unser, Cale Yarborough, Dan Gurney, A.J. Foyt, Mario Andretti, Augie Pabst, Curtis Turner, and Bobby Unser, the Ford name was welded to success in NASCAR racing for many years to come.

NASCAR and Le Mans–style racing were not to be the only entries by Ford into formal auto racing. Entries into USAC stock car, Indianapolis 500 and drag racing were forthcoming. Ford's Jacque Passino had begun development in 1961 of a special 289-cubic-inch, quad overhead cam high-performance race engine designed to run at the Indianapolis 500 track. A car powered with this engine won the race in 1965, the first win ever at Indy for a Ford-powered car. This race-configuration engine was used to compete very successfully in racing for several more years.

By early 1964 Ford was having record model year sales with sales figures exceeded only by the 1923 Model T. One of the reasons the cars were selling so well was that the intelligent businessman McNamara, who was not an automobile man, had been replaced by someone with all the business sense of McNamara together with an enthusiast's appreciation of auto design and the drive of a hard-working salesman. Iacocca was his name.

Chapter 11

MUSTANG II X-CAR

> "We took a steel prototype body, made it a convertible, took the bumpers off, restyled the front and back, and did a lot of things to pick up cues from Mustang I."
>
> Hal Sperlich

A plan was in place based on Fairlane Committee research, discovery, and management guidance. In focus was a way forward to the development of a new Ford sporty four-passenger car. Performance and styling were key, and marketing had taken care of those parameters with the "Total Performance" campaign. The swell of emotion was growing as Ford's public anxiously looked for more excitement from Ford Motor Company.

The 1962 two-seat Mustang Experimental Sports Car received rave reviews at its introduction at Watkins Glen in the fall of '62. Iacocca felt the venue would be a good testing ground for public review of a follow-up prototype that would display the culmination of the hard work of the Fairlane Committee. Important in those days of auto marketing was pre-introduction secrecy. There could be no attempt to actually display a newly contemplated four-seater sporty vehicle and show the company's cards to the public or the automotive competition before the production car would be introduced. There had to be crafted a cleverly designed vehicle which would conceptually whet the appetite of the viewing public while not revealing the true lines of the car itself.

Ford had been deluged with letters written to the company requesting a return of a sporty two-seater, like the original Thunderbird. With the showing of the two-seat Mustang, that two-seat interest was rekindled and the letters started coming in again. The Engineering and Research staff was sending out press information and individual letters in October of '62 to those people who had inquired of Ford about the new Mustang. A paragraph from the "canned" response stated, "Since it is an experimental car and represents just a first step in an engineering and styling advanced project, we cannot at this time speculate on production possibilities. The Mustang development program will continue and we will, of course, keep a close check on public reaction."

Iacocca knew he had to resist the interest in a two-seat car since the plan from day one was to make this a four-passenger vehicle. For Iacocca, the solution was simple. As Ford Division general manager, he decided the company needed a vehicle to test consumer reaction to function and styling. It was in the early spring of 1963 when the requirement for a new four-seat concept car went out and became of paramount importance to Ford stylists involved with the development of the soon-to-be-produced all-new Ford Mustang.

The X-Car prototype was built for this purpose and to test the market for a car that could sell for less than $3,000. Further, it was to provide a marketing "bridge" for the public to take them from the original two-seat Mustang prototype car to the conceptually new four-seat sporty car to be introduced by Ford.

Gene Bordinat, vice-president of Styling, asked Executive Stylist John Najjar to sketch what the Mustang II should look like. In my interview with Mr. Najjar, he told me he went home the evening of the request and sketched four 18" × 24" drawings that combined two-seat Mustang styling cues and production car prototype styling. He showed them to Bordinat the next morning. Bordinat liked what he saw and planning to proceed with the car went forward. Conceptual drawings of the X-Car were apparent by April 16 as seen in this photo.

The drawings were turned over to designer Don DeLaRossa for finishing touches. The final design was completed by May 11. Styling cues from the original Gale Halderman

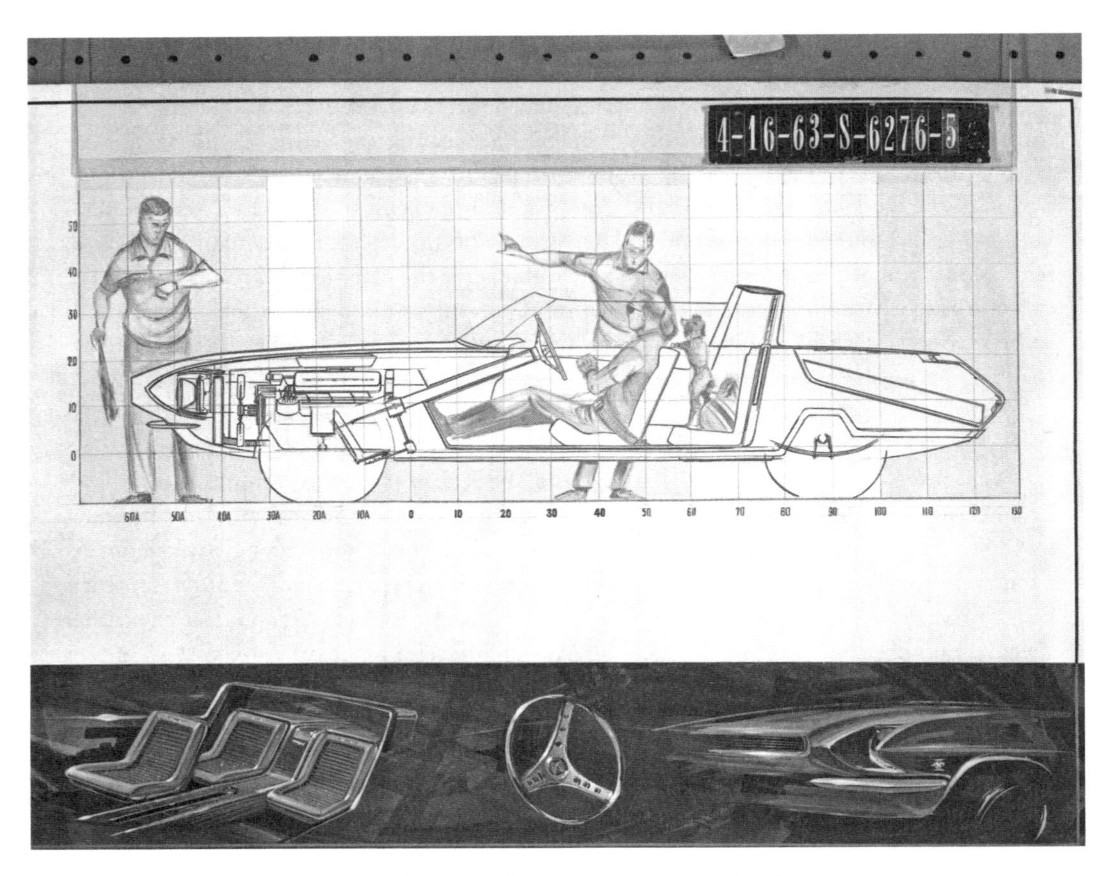

Early design drawing of the concept X-Car. *Ford Motor Co.*

1964 Ford Falcon Sprint. *Ford Motor Co.*

proposed production car sketch, such as the rear quarter mid-panel line break, may have influenced Najjar's original design. One styling cue from the original two-seater incorporated into the design was a suggested front-to-rear two-stripe paint scheme in white and blue. Once again, the '64 Falcon Sprint styling of a short nose and long trunk, which had been revised to a long nose and short trunk to pick up more of a classic look, was incorporated.

The final design of the grille and front end of the car may have been heavily influenced by the requirement for grille-surround part tooling. It was not available at the time for the production of the complicated, intricate cast pieces that would be used on the production car. A custom-made fiberglass front end fit the bill exactly, and that is what was used.

During this same time period, Don Frey had been busy with his team, and on May 2, 1963, details of the final design of the soon-to-be-built 1965 four-seat production Mustang were revealed in a design studio photograph of one of the concept cars. This photo (see page 102, top) showed a cougar emblem mounted on the grille of a fastback body concept with the name "Special Falcon" on the license plate, but on the fender was the name "Cougar Torino." Finalization of the 1965 production version continued.

Photos of the first completed clay model mockup of the X-Car were taken on May 16 at the Ford Design Center with the "Torino" nameplate on the mockup fender and a Cougar-style cat installed in the grille emblem enclosure. One day later, May 17, photos of the same mockup show the vehicle with different hubcaps and another nameplate, this time "Mustang." From these photos it is evident that the production name for the pro-

1964½ Concept with Cougar Grille, Cougar Torino name and Sp. Falcon plate. *Ford Motor Co.*

X-Car with Torino name and Cougar grille. *Ford Motor Co.*

totype X-Car was, at that time, not yet finalized. Final form for the X-Car was set and assembly of the real prototype would proceed throughout that summer.

Ford sent the final design specifications for the new "bridge car" to the prototype manufacturing subcontractor, Dearborn Steel Tubing Company, located in Dearborn, Michigan. This company was used by Ford over many years previously to produce prototype vehicles for the Ford Styling Department. Gifted designer and stylist Vince Gardner was previously chief of engineering at Studebaker and was hired to work for Dearborn Steel Tubing. He was given the task by his employer to produce this prototype car.

Ford sent a 1963½ Falcon Sprint chassis, which had been modified to 1964 Falcon Sprint specifications, over to Dearborn Steel Tubing. Measurements would be used in the project, according to my interview with Bob Negstad, Ford Mustang Project chassis engineer. This chassis was returned to Ford after initial usage and was replaced by an early prototype '65 Mustang coupe, but with the top of the coupe cut and removed to produce a convertible. It would be used as the basis for the new X-Car.

A steel birdcage was constructed on the floor pan assembly to form the basis for portions of the shell of the car to be molded from clay. The birdcage was then removed with the surrounding clay shaped body parts. Once the final clay shape was completed, a reverse mold was made to lay out fiberglass to produce the final prototype chassis components. The original steel floor and firewall, door frames, front fenders and quarter panels were retained. Steel door shells were used with exterior modifications and the interior sides were reskinned with a new fiberglass shape.

The Falcon II returned from Dearborn Steel Tubing to Ford with modifications complete. *Ford Motor Co.*

These pieced-together metal and fiberglass components were mounted permanently to the '65 Mustang floor pan assembly, producing the final prototype. Specifications called for it to be a drivable vehicle. A Ford four-barrel carbureted, high-performance 289-cubic-inch 271-horsepower V-8 engine, the same as was being used in the Cobra sports car, was installed. It was mated with a four-speed transmission. Dearborn Steel Tubing destroyed the body molds six months later as no further need was anticipated. The X-Car was returned to the Styling Department at Ford with a new nameplate on its side: "Falcon II."

Specifications called for the car to be visually outfitted just like a real car. Upon my May 1, 2003, inspection I found that none of the non-essential dashboard controls (i.e., the heater, vent, radio, lighting, etc.) were hooked up. The 120 mph speedometer was for show only and was not operational. Engine instruments were operational, including the Sun 7000 rpm tachometer. The glove box door was for show only and could not be opened. The vehicle was built without formal internal structural support in the passenger areas. I observed the front-to-rear full-length floor console was braced internally by wooden 2 × 4s. Ford advertised the car as a one-of-a-kind "fully operable steel prototype."

Don Frey, vice-president of Design and later Ford president, stated only one X-Car was ever produced by Dearborn Steel Tubing. Once it was turned over to the Ford Styling Department, Ford's design team, as often was the case, may have made some minor changes to the prototype. The car did not have any type of a data plate or identification markings on it from Dearborn Steel Tubing, according to longtime employee Jimmy "Hammer" Mason. However, it does display a metal identification tag screwed to the top of the right inner fender, apparently installed by Ford, with the marking "X 8902-SB-208*" (one engineer contacted believes X stands for "Experimental" and the remaining numerals identify the assembly drawing number for that vehicle). It also has stamped into the metal firewall-to-fender brace the number "*V5X-CONV-BEGA-X545-11-XK*." The original radiator is still installed in the car with an ID tag dated "8-26-63," which helps chronologically date the building of the chassis during August.

High performance 289-cubic-inch 271 horsepower V-8 engine. *The author*

By September 19, 1963, first photos appear of the prototype car with finished chassis as it was returned to Ford. It was a combined steel and fiberglass vehicle

Top: **October 6, 1963, press release side view photo of the new Mustang II.** ***Bottom:*** **Press release photo with a right side view.** *Ford Motor Co.*

in primer paint. By the end of the month, the chassis work would be completed and the final white paint with blue stripes, carefully remindful of the original two-seat prototype Mustang, would be applied. The "two-plus-two" interior (the first public use of this new term) was now fitted with thin-shell bucket seats in front and fully recessed bucket seats in the rear. The low profile, detachable Thunderbird-style roof design gave the car a lower and longer look, although it was only 5" longer than the yet-to-be-produced production Mustang. Final assembly was complete with the new name "Mustang II" installed. Photo sessions of the car were arranged for publicity purposes. On Sunday, October 6, 1963, formal press releases went out to the news media unveiling the new prototype.

Formally named "Mustang II," it had the original-style running horse emblems from the two-seat Mustang mounted on its fenders. Previous suggested smoke-screen names and emblems were dropped. It was now strategically named to draw attention to its 1962 Mustang lineage, yet the title cleverly designed not to draw too much attention to the new production car of the same name. The ultimate test of public reaction to the new Ford sporty car was about to begin. This was the final test by Lee Iacocca to be sure his marketing strategy for a four-seat sporty car from Ford was on target.

Prototype interior view. *Ford Motor Co.*

Wide publicity was given the new Mustang II at its first actual public appearance at the Watkins Glen Raceway in Watkins Glen, New York, on October 6, 1963. The day before the race, Lee Iacocca introduced the car to the press for the first time. He told the afternoon press conference the Mustang II was one of a series of recent show cars Ford built to test public reaction. Plans to show those vehicles to large segments of the public at auto shows and other special events would give him a pretest of likely customer response to styling and mechanical innovations that could be considered for future production models.

Little did those who saw the Mustang II know that what they were seeing was actually the new production Mustang form cobbled together with bits and pieces of styling cues to camouflage the real production Mustang shape. Even the fender "running horse" emblems were of the original two-seat Mustang design, so as not to reveal the newly proposed emblem design, which would be used later for many years on production cars.

A photo that appeared in a *Road & Track* magazine article about the car published late in 1963 clearly shows Lee Iacocca driving the car around a track with race driver Graham Hill as a passenger. Once again, the new Mustang II was overwhelmingly approved by the crowd. It soon appeared in newspapers and magazines all around the world and was also featured on CBS television.

One of the risks associated with early public showing of such a new concept was that competition would be given a head start on planning their own similar stable. There is little doubt today that the Chevrolet Camaro, Pontiac Firebird, American Motors Javelin, and the drastically restyled Plymouth Barracuda were all competitive responses to the new "Pony Car" concept by Ford.

The Mustang II was used by Ford to promote the new four-seat sporty car concept at various other coast-to-coast auto shows until January 1964. It was then displayed in Detroit at the internationally famous Detroit Auto Show. The prototype was so wildly popular that companies produced toy models, such as the Johnny Lightning "Mustang Classics" 1963 Mustang II, and the Lindberg "'63 Mustang II Original Ford Concept Car."

That Detroit Auto Show was the end of the road for the Mustang II X-Car. The "bridge car" had done its job in just 3 short months. Input from the Mustang II's acceptance confirmed Iacocca's plan for placing the new '65 Ford Mustang into production with a first-year production unit goal of 85,000 cars. The new production 1965 Ford Mustang, by January 1964, at the time of the auto show, was already well into the assembly of the final pilot cars at the Ford Allen Park Pilot Plant.

Life at the Detroit Historical Museum

It must be assumed the Mustang II was then placed in storage in February or March of 1964 somewhere at Ford Motor Company, not to be seen again by the public for over ten years. Between '64 and '67, other prototyping work was incorporated into the car at

Ford Styling by upgrading engine performance and also incorporating some minor interior and exterior restyling.

On June 17, 1974, a letter to "confirm interest in acquiring the car" was written by the Detroit Historical Museum Director to the Ford Contributions Committee expressing a strong desire to have the car donated to the museum. Apparently someone at Ford had issued a directive to dispose of the car and the museum was contacted. It was proposed by the museum the car would be used in a display titled "No Hill Too Steep," according to museum records. Ownership transfer negotiations were slow and it wasn't until one year later, on July 23, 1975, that the proposed acquisition received final approval. The museum volunteered use of their trailer to pick up the car, which Ford declined. The car showed up unannounced on the doorstep of the museum somewhere between July 23 and September 8, 1975; the exact date is unrecorded. No original documentation for the car was attached with the donation and no history of the unique prototype vehicle was passed on by Ford.

Ford required the public announcement of the donation be made to correspond with some local event of public interest so publicity "mileage" out of the donation could be maximized. By September 8, an appropriate time to announce the donation publicly was still in question. The museum suggested in a letter to Ford the donation be announced at a corporate member affair honoring the Chrysler Corporation's 50th anniversary, the Stroh Brewing Company's 125th anniversary and the Fred Sanders Company's 100th anniversary. The proposed program was not well received by Ford since the presentation might upstage the three firms being recognized. A news press release from Ford on November 3 announced the donation of the vehicle to the museum. By November 11, 1975, arrangements had been completed with the museum and a special event was created for the formal presentation in the middle of November.

The Detroit Historical Museum, as could be expected in the years to follow, was inundated with prototype car and vehicle donations from other manufacturers. The museum was unable to protect many of these cars from the elements since, being a municipally owned institution, its funds were limited. The decision was made to release on loan many of these vehicles for public display at other museums around the country that were better able to store them and provide for their long-term upkeep.

After twenty years in obscure storage at the museum, the Mustang II was moved to a new "temporary home." In the spring of 1996 an arrangement with the Owls Head Transportation Museum in Owls Head, Maine, was made for the transfer of the Mustang II. The museum maintains a large collection of various forms of early transportation, including airplanes, and the Mustang II fit well with its displays. Arrangements were made by Owls Head Museum personnel to have the car transported by common carrier to the museum.

When received, the Mustang II was in very poor condition from having been stored in a high-humidity environment over the years in Detroit. The car was dirty and rundown and was not in running condition. The Owls Head Museum cleaned and minimally restored the car to its present-display condition. No attempt was made at that time to return the vehicle to good running condition.

The Petersen Automotive Museum in Los Angeles, California, asked my assistance in producing a "40th Anniversary of the Mustang" display for presentation at their museum in September of 2004. I was able to gain approval for use of the car and I arranged transportation of the Mustang II from the Owls Head Museum to Los Angeles, where it was placed on display with the first pre-production Mustang coupe, and two other later model Mustangs representative of the era. I asked Mr. Iacocca, who resides in the area, to appear at a press conference with the cars at the Petersen. He graciously agreed, and was reunited with his creation, the Mustang II, after 41 years. For the first and only time the Mustang II prototype was photographed alongside the first pre-production coupe with Mr. Iacocca. It was at the display I first learned the prototype actually would run and drive, although not well, after sitting for all those years, mechanically unattended.

As Iacocca reminisced about the car, he said, "I remember it well. This was the first prototype that approached being a real car out of all the previous prototypes Ford had built over the years. This one actually ran and it drove really well. I drove this car, you know. I remember driving it in 1963 when I took race car driver Graham Hill with me for a spin around the track. I never drove it again. It had lots of power from that Cobra engine we had in it." As he further reminisced, "The car had a number of Thunderbird influences designed into it. Gene Bordinat was involved in the styling. I had the car made

Lee Iacocca reunited for the first time with the 1963 Mustang II on display at the Petersen Automotive Museum in 2004. *The author*

for marketing research — and the people loved it! We sent a '64 Falcon floor pan over to have it built. This is the only one we built. We had the front and rear ends really modified, as you can see. We sent it around the country for public reaction." I felt Mr. Iacocca thoroughly enjoyed the togetherness of the two historic cars as his eyes reflected excitement. He was able to relive a thoroughly challenging and enjoyable time from his career. At the end of the Peterson exhibit in the fall of 2004, the Mustang II was returned to the Maine museum.

On March 6, 2009, it was loaned to the America on Wheels Museum in Allentown, Pennsylvania, for display until September 30, 2010, for all those who were interested in seeing an integral piece of Mustang history. The Detroit Historical Museum retains ownership of the vehicle.

The Mustang II, over the years since its introduction, has been called the most beautiful Mustang ever built. The original concept was for the sculptured steel shape to simply

The 1962 two-seat experimental Mustang compared with the 1963 four-seat prototype Mustang II. *Ford Motor Co.*

Farewell Mustang II. *Ford Motor Co.*

be a "public teaser." It did that. It was truly a one-of-a-kind car made to "bridge the gap" for Ford marketing from a two-seat prototype to a four-seat production reality. It did that, too!

It now has passed into motoring history — its job accomplished, and accomplished well indeed. Its name Mustang was retired and would never again be used by Ford relating to a concept vehicle.

The Only Mustang Prototype Vehicle Known to Survive

The Mustang II was built during the late spring and summer of 1963, months after the final design of the 1965 production car was established in September of 1962. Even though it is a one-of-a-kind true concept car built on a prototype chassis, it is not the first four-seat Mustang assembled. In the seemingly never-ending quest for identifying the "first ever" Mustang assembled, there were production-style prototypes already built for testing when it was decided to produce this car. This chassis was one of those first prototypes assembled, but its creation came after the first prototypes were produced in April 1963. Assuming its use chronologically in the prototype build cycle, it was most probably not "the first prototype built."

The Mustang II is the only known surviving prototype, even though highly modified, of the production 1965 Mustang. It shares not only true four-seat design and close-to-

Iacocca and Bob Fria discuss the Mustang II at the Petersen Museum. *The author*

production exterior styling, but also an almost exact full floor pan, firewall, engine and trunk compartments, as well as engine and drive train components.

History will show the Mustang II was an important part of the development and marketing of the real production Mustang, more than the original two-seat version could have ever been.

Mustang II Specifications

(1963 Mustang II compared to 1965 Mustang)

	1963	*1965*		*1963*	*1965*
Chassis					
Wheelbase	108.0"	108.0"	Overall Width	68.2"	68.2"
Overall Length	186.6"	181.6"	Front Tread	56.0"	56.0"
Overall Height	48.4"	51.0"	Rear Tread	56.0"	56.0"

Engine & Transmission

Ford	High Performance 289-cubic-inch four-barrel V-8 with 271 horsepower
Ford	Manual 4-speed transmission

Mustang II

(Text quoted from Ford Public Relations brochure — undated)

Created to test consumer reaction to certain design and engineering innovations, the Mustang II gained such favorable response that Ford Motor Company eventually introduced a somewhat similar vehicle as the 1965 Mustang production car.

Evolved from the Mustang I, which is a two-seater open sports car, the Mustang II has a "two-plus-two" configuration. This means comfortable space for two front-seat passengers, and room for two additional passengers in the rear seat.

A fully operational steel prototype, the one-of-a-kind Mustang II is powered by a 271 hp V-8 engine.

Pointed front fenders and simulated air scoops faired into the rear quarter panels of the Mustang II are reminiscent of the original Mustang's design. The experimental unit has a detachable hardtop roof designed to be a refinement of the original Thunderbird roof.

Headlamp coverings are faired into the nose of the car to preserve the aerodynamic appearance of the front end. The grille air intake thrusts ahead of the front fenders and bears the Mustang emblem in a frame supported by chrome crosshairs.

The interior of the Mustang II has molded, thin-shell bucket seats both front and rear. The seatbacks in the covered rear compartment are integrated with the rear deck. A console, also suggestive of the Thunderbird, sweeps up to merge with the instrument panel.

Chapter 12

"Now, Mr. Ford ... Listen!"

"There was nothing magic about it. All of the stars were in perfect alignment."

Lee Iacocca

By late 1961, Iacocca's informal team had envisioned a spectacular introduction for their new Mustang. What better place to introduce the car to the world, as a mid-year car model, than at the 1964 World's Fair? They had set the timeline many moons earlier to coincide with the fair's opening in April 1964. The team knew automotive media attention to the Mustang would have grand-scale focus.

For the first time in 403 years, on February 2, 1962, the planets Neptune and Pluto came into perfect alignment with our Earth. Maybe this was the precursor for the kind of year 1962 would be for Lee Iacocca and the Ford Motor Company.

In the spring, new ideas were being mined like gold at the corporate headquarters. A banner year in Ford innovation had begun. Development of the '63 model year cars was well on its way, and Fairlane Committee concept vehicles were evolving.

Planning and building four-passenger sporty prototype concepts had by now become more formalized. A myriad of examples were flowing out of the styling department by the end of 1961 and into early 1962.

Designers deep within Gene Bordinat's styling department were busy finalizing plans for the new concept two-seat experimental sports car to become named Mustang. By May, formal go-ahead plans were signed and by June 1, the team had a mere 100 days to build that two-seat car.

Gale Halderman, working in the Oros department on the four-seat concept, said one goal was to make that car a "personal car," something a customer might buy just for going to work or the grocery store, and not necessarily a car to be used as a family or weekend type vehicle. "We'd never done a car that did not have a full rear seat. This car would have a compromised rear seat, and designing that was a major concern." That portion of the design was directed by project manager Hal Sperlich.

Early in 1962, Sperlich presented some paper sketch designs for the concept car to Henry II, who promptly rejected the whole idea of building such a car. Regardless, the

***Opposite top*—Eugene Bordinat (on the right) with Henry Ford II.** *Ford Motor Co.* ***Bottom*— Gale Halderman.** *Courtesy of Gale Halderman*

team was prodded on by Iacocca, and by mid-year, the first formal program proposal to build the sporty car, known as the T-5 Project, was presented at the corporate product approval meeting in the form of what was known as a "Blue Letter." It was promptly rejected. Not only was it rejected, but in the middle of the meeting, Henry jumped up, said "I'm leaving," and promptly left the room. Iacocca was crushed. Unbeknownst to Iacocca was that Henry had been feeling ill, became sick and spent the next six weeks in bed with mononucleosis. Apparently the bed stay gave him time to think about Iacocca's proposal, because he came back after recovery much more receptive to the program.

It had been a busy early summer for the country. On June 25, the U.S. Supreme Court ruled mandatory prayer in public schools was unconstitutional. A week later the first communications satellite was placed in orbit by AT&T. Of note, on the 4th of July, future actor Tom Cruise was born. Viewpoints were changing, and so it would also be at Ford.

Iacocca was restless with all the indecision within Ford about the new car. His instincts told him he had to be settled down on a close-to-final design quickly in order for the entire production and marketing strategy to come together by April '64. "I had a very talented team of designers as a group put together by Gene Bordinat upon inspiration from his boss George Walker," Iacocca told me. "I knew I could count on the team to come up with what I wanted."

Styling boss Bordinat was himself anxious and chided project manager Don Frey, "We spent nine months trying to figure out how much room the back seat should have, and we've ended up with a design that is just about what the prototype Allegro is dimensionally." And not approved.

By July 27, Iacocca issued his edict to Bordinat to have a design competition between three styling departments to produce a final version of the car that would be acceptable to all. Hal Sperlich stated in our 2007 interview, "There were actually seven teams of stylists involved." Although there was no money budgeted by the Ford Division for production of these clays, Iacocca found the money and ways to funnel it to the project and see it through to completion.

Ford Studio manager Joe Oros happened to be attending a ten-day conference off campus when the Bordinat edict for a new design was issued. Dave Ash was the Ford Studio executive designer, and in Oros's absence, took it upon himself to immediately start work on a clay model with his choice of design. He worked the design across the hall from the Ford Studio in the small Advanced Studio. It was to be done by the time Oros returned. That particular design was reportedly very stiff and boxy. It had a reverse-angle rear window, similar to the '59 Lincoln design. Upon Oros's return, he immediately stopped work on that model and decided a new design had to be selected. He wanted more of a "snarky nose" design incorporated with a rear wheel forward side air scoop, according to model designer George Schumaker. Ash would call for new designs from his team headed by design manager Charlie Phaneuf so the "young guys could get involved," said team member Schumaker. The team worked around the clock over that weekend to produce new sketches.

Oros was desperate as he only had about 10 days to design and finish a full-size clay model. Design Manager Gale Halderman was heavily involved with the design work for the new '65 Fords in the Advanced Studio when Oros brought him to his Ford Studio to work on this project. Halderman protested, not having the time to work on yet another project. Oros insisted, and Halderman was handed the additional assignment to style yet another model car for the Iacocca competition.

The only time Halderman had to work on the design was at home, as he was busy at work running back and forth between studios to check on the progress of the '65 Ford car designs. "The five sketches I did on my kitchen table at 10:30 one weekend night were posted in the morning for the 8 am meeting on the wall of the Styling Studio along with those of the Dave Ash team and Joe Oros." Three of the Halderman sketches relied on the side air scoop in the quarter panel and two used the hop-up style quarter panel. Halderman states, "I think having three sketches with similar characteristics made a strong case." Oros was generally not in agreement with typical Ash-directed designs as the two didn't see eye-to-eye on many styling issues, according to Halderman.

As the sketches were reviewed by the three men, Oros decided on the one that Halderman had done and decreed, "That's the one we're going to build," Halderman told me. "We proceeded to make the first clay model. It relied heavily on the scoops on the side and the hop-up quarter lines. The front end was primarily designed afterward. Oros had me work on the taillamps, which I designed as rectangular and no longer of the traditional round taillight style, to give the design a newer and more appropriate crispness. I also worked on the rear end design which I did in one night while Oros and designer Charlie Phaneuf did the front end." Some of the Ash styling was also included in the chosen design, including oval Ford Taunus headlamps.

Most of the subsequent changes made to the original Halderman design sketch were made for engineering to accommodate the Falcon chassis control points and seating package. The result was a higher roof line yielding less freedom in the original front and rear design.

Halderman styling was applied to the left side of the styling clay form. Joe Oros, with modeler George Schumaker, created a design that was applied to the right side of the same clay. This produced a final clay model that was asymmetrical, left compared to right, thereby eliminating the need for two separate models.

Iacocca visited the studio with Frey and Sperlich and they were shown the car's preliminary Italian-influenced styling of the front bumper and grille as enhanced by the Halderman-designed left side. Oros said, "He really brightened up.... When he got excited, his cigar started twirling. And it started twirling." The entire triumvirate agreed on the looks of the Halderman-styled left side. Halderman relates, "Lee said, 'This is it! Call Henry and tell him to get over here right away.' Henry arrived within the hour and after reviewing the car agreed it was a great design. Then he said, 'Go ahead with the design, but I'm telling you right now it's not approved, but I'm not telling you to stop.' With that, he left. Iacocca turned to the designers and said, 'We've got it!'"

On August 16, the final six prototype clay designs were assembled in the Styling

Top—The left side format of the August 16th double sided styled clay with oval headlamps called Cougar was selected for production as the new Mustang on September 10, 1962. Fairlane Committee planning led to the creation of this first production approved version. ***Bottom***—The clay right side styling was not chosen as the design for the production Mustang. *Ford Motor Co.*

Center courtyard for viewing. The Halderman design produced in the Oros department and incorporating some of the Dave Ash team influences was selected as winner of the styling competition. It was named Cougar.

For reference, at that same time, in Los Angeles, the 1962 two-seat Mustang prototype body was in its final assembly stages at the Troutman and Barnes facility. It would be completed in early October for its first showing at Watkins Glen, New York.

By now, the second and the third "Blue Letters" for the new four-seat platform had been presented for corporate product approval and turned down. The approval committee negativity was generated by three serious concerns. The committee was not convinced the newly defined youth market was real. Also, the recently disastrous Edsel fiasco had cost the company a fortune and it was not about to commit millions to a new concept model car. Third, it had already committed $250 million to the Ford Division alone for the new 1965 Ford model retooling. These were almost insurmountable odds in trying to sell the new Cougar, as it was then named, to the committee.

However, Iacocca was successful in drumming up enough interest that committee member, and soon-to-be-installed Ford president Arjay Miller (1963–68) ordered a study of the proposal. Miller was one of the original Whiz Kids hired along with McNamara. He was concerned the new model might cut into sales of the Falcon, with the net financial result being a loss overall to Ford. The study came back with results showing they would have to sell 86,400 cars the first year to break even. Don Frey said years after the Mustang success, "We had no research to come up with that number, so his researcher just made the figure up. I felt they didn't need further market research, we just knew what we were doing."

Time was getting short if the original plan to present the new car to the public at the World's Fair in April '64 was going to be possible due to development and production lead times. Normal time to get a new model from approval to production was about three years. Feasibility studies of where the car would be built, how much it would cost, and in fact whether it could even *be* built had to be completed for committee approval. The team had eighteen months to make the car a reality.

Iacocca was convinced the styling for the new car was perfect. There was also one more concept that the Sperlich, Oros, Halderman and Iacocca team had agreed on: the car had to be sold as a complete automobile. "We couldn't sell it as a base model and have a stripped-down model and then end up charging more for carpeting, more for wheel covers, more for good seat trim. This car was done with wheel covers as a base and with full carpeting and we offered it as a convertible; it was a full line of car. It was the very first time this was ever done at Ford so it was a new era of car design," said Iacocca. "The car would have its own personality: demure enough for church going, racy enough for the drag strip, and modish enough for the country club."

Henry II was not yet on board to support recommending this new production model. Iacocca took him to the styling center courtyard for another serious look at the clay model. Gale Halderman, who was present, is quoted as saying later, "When Mr. Ford saw the clay, he was enthused by the looks of it. I think its appearance is what inspired [the eventual completion] of the whole project." Iacocca told Peter Collier & David

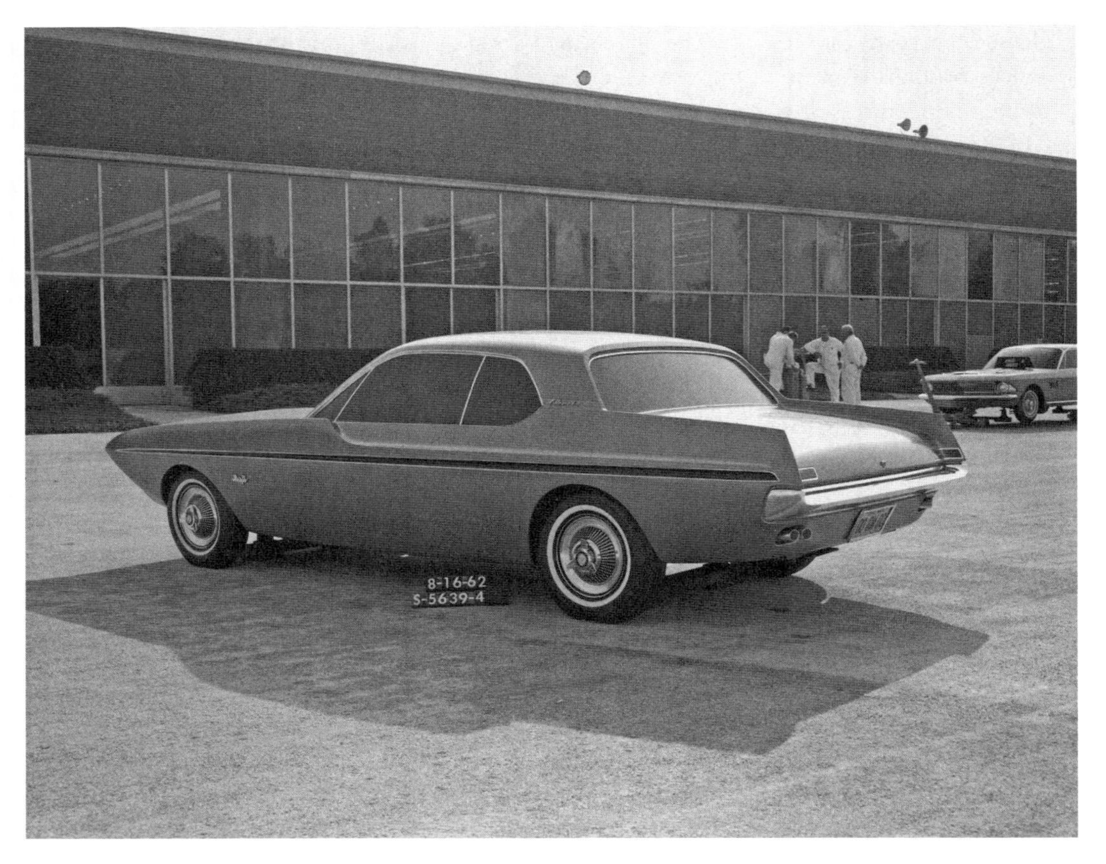

August 16 Lincoln-Mercury Studio competition entry clay not selected for production as the new Mustang. *Ford Motor Co.*

Horowitz, authors of *The Fords*, that he recalls Mr. Ford verbalized that day, "Enough of this shit, Lee. We got our ass lost on the last one. Who needs another?"

That "last one" was, of course, the Edsel, on which Ford Motor Company had taken a $250 million loss. The loss was obviously still fresh in Ford's mind; as Iacocca related to this writer, "He chastised me and called me an E-guy. Mr. Ford referred to anyone who had worked on the Edsel project an E-guy and constantly reminded them that their career with the company was tenuous. I told Ford that I was *not* an E-guy."

Iacocca knew that he had to have a clay model completed and approved by September 1 in order to make the World's Fair deadline and Ford was yet to be sold on the idea. So once again Iacocca traveled from his fifth-floor office in a building that was about a mile away from the headquarters, rode the executive elevator to the 12th floor, and knocked on Ford's office door. It was protocol; one always knocked on the door and one always addressed him as "Mr. Ford"—or else.

Iacocca was known to be a tough-talking and demanding executive who could be very blunt, and Henry liked that. It was time to be blunt. He was there, again, to gain

Henry's approval to build the car. His requests were becoming old and tiring to Henry. As he addressed Mr. Ford, his boss began to see the general manager had put the obligatory effort into the development of the project and realized he wasn't apt to get the aggressive Iacocca to back off. He jumped up and said, "I'll approve the damn thing. But once I approve it, you've got to sell it, and it's your ass if you don't." (There are several versions of what was actually said that day in Ford's office, but this version seems to be the most nearly correct.)

On September 10, 1962, there was another corporate product approval meeting with a new "Blue Letter" written by Hal Sperlich. Iacocca would make a brilliant pitch to the committee while Sperlich ran the slide projector. This time Henry II agreed to support the plan to build the new car, but with one change to the design as it was presented by Don Frey, Gene Bordinat, Hal Sperlich and Lee Iacocca. He wanted one inch more in the back seat. Iacocca told me, "He got it. I had to agree to it just to get him to approve the plan." Henry, to save face, reiterated at the meeting, "Okay, I'll approve the damn thing just to get you guys off my back. But it's your ass if it doesn't sell!"

And so it was on that September day, Iacocca gained reluctant corporate approval in writing for Project T-5. He had a check for a modest $45 million, not the $75 million he had requested, to tool and develop his new car called the Cougar. At that time, normal new car development costs were $300–$400 million, substantially more than what was just approved. Henry set a target date of March 9, 1964, for Job One and the first production cars to drive off the assembly line, a mere five weeks before the scheduled public introduction in time for the World's Fair. The consummate salesman Iacocca had just been through "the toughest selling job of his life." He had just sold the reluctant managers of the world's second largest car maker a new unproven concept car. Little did they know what the future would hold. Job well done, he drove home to Bloomfield Hills where Mary, four-year-old Kathryn and a little black schnauzer named Mr. BoBo were waiting.

Two days later, on September 12, President Kennedy, in a now-famous speech, would issue the challenge to the country to put a man on the moon by the end of the decade. That very upbeat mood of the country was changed overnight to a more somber note on October 14, when President Kennedy would reveal the U.S. detection of Soviet offensive missiles being installed on the island of Cuba and aimed at the U.S. The reality of the threat of a global nuclear war resulted in an international confrontation known as the Cuban Missile Crisis.

This was less than two weeks after the '62 two-seat Mustang prototype was finished and approved by Gene Bordinat and Don Frey on October 2. It was loaded into its transporter for the trip to New York to be ready for display on October 7 at Watkins Glen. Remember, the four-seat Mustang as we know it had already been approved for production just weeks before the two-seater would be shown for the first time, further proof the 1962 version had little to do with the final Mustang design.

On November 9, tragedy struck Ford Motor Company when the Ford Rotunda building burned to the ground. Destroyed building contents included decades of recorded Ford history. The coming year of 1963 would hopefully start out better, with Ford Motor Company producing its sixty-millionth vehicle.

At home, Mary, Lee and Kathy (age 3) Iacocca. *Walter P. Reuther Library, Wayne State University*

In the spring of '63, with John F. Kennedy as president and the youthful atmosphere that came with his administration, Lee Iacocca was general manager of the Ford Division at 38 years old. His wife Mary was suffering from the effects of diabetes while raising their daughter Kathy, now age 4. He was born of an Italian immigrant family in the Lehigh Valley of Pennsylvania and was taught to be a Catholic with strong family values. No matter what happened at the office, the priority was always with family, so much so that he would cut short executive meetings to be present at school events for his kids. His father had always told him, "Lee, you have the freedom to become anything you want to be ... if you want it bad enough and you are willing to work for it." With strong allegiance

Final body design wearing Cougar name. *Ford Motor Co.*

to his father's watchwords, he marched to his own beat in his ascendancy at Ford. Even as a kid, he always wanted a career at the Ford Motor Company, and let nothing stand in his way. He states in his book *Iacocca: An Autobiography,* "Whenever I've taken risks, it's been after satisfying myself that the research and the market studies supported my instincts. I may act on my intuition — but only if my hunches are supported by the facts." He was about to embark upon a course at Ford that would change his career and his personal life forever.

The grand scheme of Fairlane Committee planning was now moving rapidly into advanced development of the marketing program. Iacocca needed a sales tool that could gauge the pent-up youth interest in his new car. Thus was born the idea of the Mustang II prototype.

On May 1, 1963, the recently selected new Ford Motor Company President Arjay Miller was formally confirmed to that position by the board of directors. Iacocca and his team would get a new boss. On that same day, the final design plans for the new four-seat Special Falcon, as it was called, were cast firm for the Ford Division. Engineering and Planning Departments had already begun the procurement of the parts needed to produce the pilot cars.

By mid–May, things were going very well indeed, with the best yet to come.

Chapter 13

PREPARE TO LAUNCH

"It had the excitement of wide-open spaces and was American as all hell."
J. Walter Thompson Ad Agency

Every new vehicle must have a recognizable name and, in most cases, an emblem that symbolizes the car. Our car needed both at this point.

During the original T-5 project presentation the car was first referred to as "Special Falcon." As an abundance of concepts were developed, names appeared such as XT-Bird, Avanti, Median, Avventura, Stilleto and Allegro. Henry II chimed in with a request the car be named the T-Bird II, but no one else seemed to think that was a good choice and that name was deleted.

After many concepts down the road, Joe Oros and Dave Ash reportedly saw their own new creation as feline in nature, so they gave it a name they both agreed upon, "Cougar." A stylized cat emblem was centrally located in the middle of the grille ornamentation. This is the name that was given build approval by the corporate Product Approval Committee on September 10, 1962. Iacocca knew that coming up with a name for a car is almost as difficult as designing the car itself. A name choice is subjective and can be an emotional thing for those involved with the selection.

Time was running short in the fall of '62, and the list was narrowed down to four names: Monte Carlo, Monaco, Torino and Cougar. Checking on name availability with the Automobile Manufacturers Association, it was discovered the names Monte Carlo and Monaco had already been reserved by other manufacturers. This left Torino and Cougar. The name Torino, which is the name of a city in Italy, was chosen to assist in retaining some of the car's original Italian-inspired styling cues. The Cougar emblem would be kept in the grille as a compromise.

When preparing the ad campaign for the Torino, the public relations office called Iacocca and reminded him Henry II was having an affair with an Italian beauty and a divorce appeared imminent. The car's connection to an Italian-related name could, at that time, cause Ford unfavorable publicity. That name was dropped.

Even as late as July 29, 1963, Process Engineering Department documents show the car was still referred to as the '64½ Special Falcon. The time for name selection was very short; they had to come up with a new final name immediately. Typically, Ford used automobile names that had been proposed by its ad agency. They turned for help to John Conley (deceased 2007), a name specialist who worked for J. Walter Thompson, the Ford

Concept car with Cougar grille and Torino name. *Ford Motor Co.*

ad agency. He was dispatched to the Detroit Public Library to look for names specifically related to animals, ranging from Aardvark to Zebra.

Conley's list had many possible name suggestions until it was narrowed down to six names: Bronco, Puma, Cheetah, Colt, Mustang and Cougar. Oros and Ash, the original concept designers, pleaded with Iacocca to keep the original name Cougar. That was not to be. Iacocca and Bordinat were in agreement that the name Mustang would become the new name for the car as it suggested "moving fast through the countryside." So, in early November 1962, the name Mustang was finally and officially assigned to the car.

Iacocca and Bordinat proceeded to swear Dave Ash and John Najjar to Mustang name secrecy during the development of the symbology to be used. They worked to design the new logos which would replace the Cougar emblems. Oros was not pleased when he found out the name was changed, but decided to just accept it. The Cougar and various versions of Mustang emblems were used off and on with different prototype models. Photos as late as mid–January '64 show variations of the horse emblem in the grille enclosure. (Iacocca related to me that he still has the Cougar grille emblem that was boxed by Joe Oros and given to him with a plea letter to use the name Cougar.)

Then a new problem surfaced: which way would the symbolic pony face when installed, to the right or left? The young stylist Phil Clark had designed the original '62 Mustang emblem with the pony facing left. Since childhood, he had been infatuated with horses. He dreamed of having his own car line with the name "Mustang Stallion," according to his daughter Holly Clark in her book *Finding My Father*. His drawings were typically made with the horse facing left, which to him was a more natural position to draw because he was right-handed and it seemed to be the more eye-pleasing view. All his prototype design work for the emblem was with the pony facing left, and that's the way it was presented to management.

For those opposite-direction proponents, their observation was that as most people typically thought of horses, they thought of race horses, which ran to the right from the starting gate, so the pony should face right. There were some drawings made with the pony facing to the right. Iacocca pointed out that "the Mustang is a wild horse, not a domesticated racer," and no matter which way it ran, he felt sure it would be headed in the right direction. It was decided by Gene Bordinat that the Mustang pony will always be viewed as running to the left, as Clark had designed it.

Not many people today, even Mustang collector aficionados, realize the production '65 Mustang pony emblem is different from the original '62 prototype Mustang emblem. Although they look the same from a distance, they are very different. Lee Iacocca told me in 2006 that even he didn't realize the emblem was different. As can be seen in the adjacent photograph, with the two emblems compared together, the original 1962–63 two-piece horse emblem had more of a galloping stance with a different configuration tail.

The later one-piece '65 production pony emblem portrays the horse in more of a steady run with a more stretched-out, flowing tail. It is the one that began appearing on prototype cars as they were built in early '63. Gale Halderman claimed in our interview he and Joe Oros were the designers of the new running horse, created to resolve a technical problem. The '62 style horse with the extended galloping legs wouldn't fit in the allotted space of the grille emblem enclosure. They redesigned it to a more flowing horizontal design for the purpose of making it fit into the already-styled enclosure that became known as the "corral."

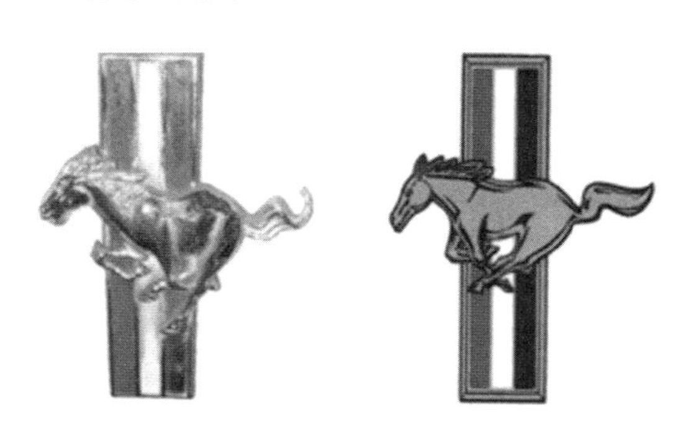

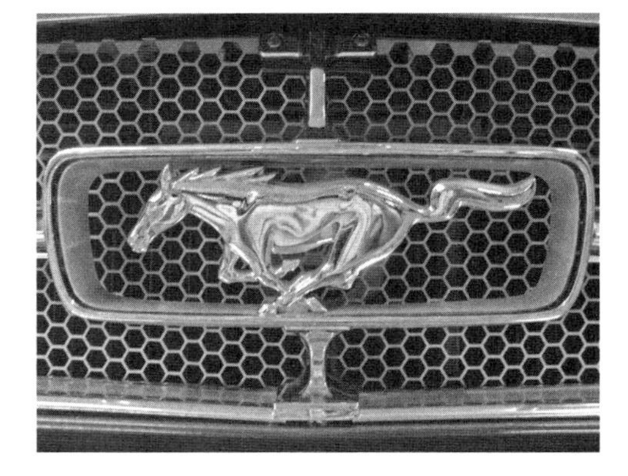

Left—**Mustang emblem evolution revealed in a side-by-side comparison. Pony legs were restyled to fit the emblem into grille corral space on the '65 grille.** *Ford Motor Co. and the author.* ***Right***—**1965 Mustang grille corral and running horse.** *The author*

There had to be two separate renditions of the pony brought to three-dimensional life to produce the final styled emblems. One would be used to make the corral emblem and one would be used in making the fender emblem with the three vertical bars, a version which required less 3-D relief. As Gale Halderman recalls, two talented artists were used within the studio to create the 3-D horse. It was brought to life by a tall and talented Hungarian sculptor with an aristocratic background named Charles Keresztes. He possessed an advanced degree in sculpture.

Keresztes claimed to be a member of the Hungarian Olympic equestrian team at the start of World War II. He joined the Hungarian military as a member of the cavalry and became engaged in the war. During battle, his commander was killed and he assumed command of his unit. He led his troops in the last horse cavalry charge in the history of the world at age 19.

He later emigrated to the United States and was hired by Ford Motor Company. As a styling department sculptor at Ford, he is credited with bringing one of the two required horse renditions to life by creating the 3-D relief of the Mustang horse. How appropriate! Later he sculpted the Cougar head for the Cougar cars.

The other required relief was brought to life in wood by modeler Wayno Kangas, a long-time modeler at Ford. He reportedly possessed superior sculpting skills, and years before had been chosen by Henry Ford to carve wooden toys for the Ford grandchildren, including Henry II. Wooden Mustang horse models he carved for the Mustang project were covered with aluminum foil and were mounted on prototypes.

The name for the car and the emblem badging had been finalized. The emblem badging was in final form and utilized the running horse with the three vertical colored bars. It was time for the first metal to be cut.

The decision to proceed with production of the new car was predicated on using the existing assembly line at the Dearborn Assembly Plant (DAP), commonly known as the "River Rouge" plant. This was the same place the historic Ford Model T was assembled.

Fiberglass mold used to create the die to cast the Mustang horse grille emblem with Charles Keresztes' initials engraved on rear side (9" × 18"). *Courtesy Eric Schumaker*

The car was to be built on the same assembly line as the Ford Fairlane car, so the two car lines were integrated together to form one line. Economically, this was beneficial since many of the parts used on the Mustang were of Fairlane derivation, and were immediately available.

By September 21, the first

semi-finalized drawings of a convertible model were shown. This is the earliest dated drawing known of the convertible, even though plans were to produce this model at the time the project was approved for production.

Fall 1962

September was the first time the Ford Engineering Department got involved with the car. This was unusual because engineers were usually called in earlier in a car's development to help solve engineering-related problems. To keep styling-related issues intact, engineering was purposely not brought in until the last minute. Clay modeler Jim Ruth had finished his final clay on the wheel cover design and modeler Don Carr had snapped the hood crease alignment string for its last time. Final styling changes were complete.

Ultimately, engineering required very few changes be made to the original design. They worked hand-in-hand with the styling department to retain the original design features. As an example, one problem for engineering was that the front lower valence panel was raked too far under to clear chassis structural members. To retain styling cues, engineering skived off the structural members to allow for clearance of the panel. Conceptually, the new car was only a body engineering job, as the basic chassis, engine, suspension, and driveline were, by design, off-the-shelf Falcon and Fairlane components.

Ford had an in-house styling rule book which outlined "dos and don'ts" of production

Special Falcon convertible concept drawing. *Ford Motor Co.*

body engineering. Certain problem resolutions were to be adhered to, such as minimal bumper to sheet metal clearances. The use of die-cast headlight bezels was prohibited and radical rear fender tuck-under was defined. Eventually, engineering "bent" 78 different rules to build the car the way styling wanted it built. Somehow, this was all okay with those involved, and it was done in the spirit of camaraderie just to get the job done on time.

Because they were "re-skinning" an existing Falcon platform, all the various chassis components were attached to the underside and all of the body components were attached to the topside. The chassis was to be beefed up to contemplate the installation of all the different engine combinations. The convertible would use heavier gauge steel and extra reinforcements in the rocker panel areas. Engineering began calling for changes to Falcon and Fairlane parts that could be modified slightly to also work on the new car, such as a countersunk filler cap neck on the Falcon radiator. Many new parts were assigned dual usage on these car lines and would be available months before the first new car was assembled.

When final engineering was complete for the car, it would be 1½ inches longer than the original design and 108 lbs. heavier, but it could still be built at a retail price of $2,368, which was under the target of $2,500 laid out by the Fairlane Committee.

In November, the new model was still being referred to as the T-5 and the Special Falcon, even though the pony emblem symbology had been finalized. It was decided to build only hardtop and convertible models. As the year ended, planning was beginning to turn into finalized material orders as the requirement for steel body molds was forthcoming.

Spirit was high in the program. It is a great tribute to those involved that an entirely new model could be pushed through to production in record time, because of the high degree of professionalism, attitude and work ethic displayed by those involved. This message came through loud and clear with all of those I interviewed for this book.

1963

The beginning of 1963 brought new feelings into the life of Henry II. Things seemed to be clicking along quite well at the shop and he established a long-term relationship with his new lady friend, Cristina Vetorre Austin. He felt so upbeat he bought a 109-ft. yacht for $700,000 which he named the *Santa Maria*. The name seemed to symbolize that he was cruising off to a new direction in his life, although he never admitted to that.

Brave expressions were flowing from the styling department on model design. A model was created to show how the new Mustang would look with four doors, a concept that never became reality.

A convertible prototype was displayed in early February with fog lights, a corralled lion designed by George Schumaker on the grille, and a "Special Falcon" license display.

The proposed hardtop and convertible were the chosen styles to be produced; how-

January proposal of a four-door named Falcon. *Ford Motor Co.*

Special Falcon convertible concept. *Ford Motor Co.*

ever, in March of '63, designs for a fastback model were shown in a clay model in the styling department from first sketches made in December.

This was more in accord with the roofline of the Allegro prototypes, even possibly influenced by the successful introduction of the new '63½ Galaxie fastback model lines. By the end of March, a clay of the proposed fastback model was shown. Halderman reports, "I wanted the fastback roofline to go all the way to the rear of the car, but Oros decided he wanted the roofline to terminate at the trunk deck line. That's how that styling issue was decided." Modeler George Schumaker takes credit for that final roof design.

A completed concept clay, colored candy-apple red, was placed in the styling courtyard and covered with a tarp. Iacocca came one day to visit the styling courtyard and inquired, "What's under the tarp?" The tarp was removed and Iacocca saw the fastback concept for the first time. His cigar twirled and without hesitation he said, "Let's build it," according to my Halderman interview.

The styling department was overloaded with other model design work and needed help completing the final design of the fastback model. Andy Hotton, owner of Dearborn Steel Tubing, the prototyping contractor used by Ford, was asked to design and present a fastback creation resolving the complex roof design issues. A roughly cobbled-together

Mustang fastback concept March 21 styling clay. *Ford Motor Co.*

Mustang hardtop mule (test bed) was given to Hotton to use for development of a prototype. By August, he had created an all-metal chassis. He used Kirksite tooling to press out the several desired metal roof panels which were then welded together and fitted to the chassis.

The styling group liked the design, and it would become the third body style in the production lineup. The decision was made to put aside further work on a fastback model for the time being due to numerous changes required for interior styling. All efforts were then concentrated on the hardtop and convertible models. These would be the only two models offered by Ford in the original April 1965 Mustang line.

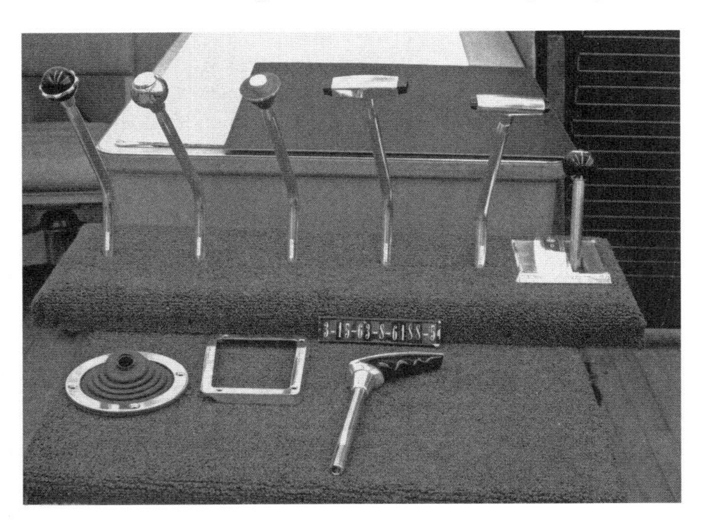

Available selection of shift levers to be decided upon. *Ford Motor Co.*

Prototype Styled Steel Wheel proposed in November 1963. *Ford Motor Co.*

By March, details of prototype Mustang-specific parts were being displayed for the decision makers to select final designs. A number of differently styled gear shift levers, for example, were shown.

Upholstery patterns and materials were to be selected, exterior light lens designs were yet undecided, etc. The original three separate vertical bar taillights were incorporated into a single light bezel to satisfy accountants concerned with the extra cost of two more light bulbs. Additionally, stylists and engineers couldn't agree on the way to build the high, "mouthy" grille. Iacocca wanted the mouth to come out forward of the fender line so the car had a long, expensive sports-car look. Gene Bordinat recalled, "It was a tough car to put together with that high mouthy grille because we didn't have plastic materials to work with in those days. There were a lot of joint matching problems with the cast metal pieces." Stylists refused to bend and over the next three months, several different versions of the "mouth" and its grille were shown. In the end, that "snarky" extended mouth design Joe Oros wanted was retained, and is distinctive in the overall design.

Stylist John Najjar's innovation of stamping "soft cow-hide grained imprints" into the metal door interior stampings had to be integrated with interior styling for three trim levels: base, sporty and luxury. Vinyl tops for the hardtop had to be designed and materials selected. Optional items, such as steering column–mounted "Rally Pacs" with a clock and tachometer, were designed. Selection of instrument panel components was necessary to provide proper in-dash display. Engineering and production matters were keeping Ford employees assigned to new car design very busy.

Other divisions such as engine, and transmission and axle, were slow to participate in the prototyping project because there was no money spelled out in the budget for them. Once they realized this car was going to be a winner, it seemed overnight everyone wanted to become involved. Adapting Falcon components to the car chassis involved repositioning brackets and attachments. Suspension and steering would be modified from Falcon parts to the more substantial Fairlane and heavier-duty parts.

Modifying the floor pan with a higher transmission tunnel was required because the car sat lower and the required bucket seats designed by Johnnie Najjar, then head of interior design, would fit better. He transferred his design of those new bucket seats from the '64 Thunderbird styling. When the first prototype was assembled, workers found it went together more easily and with fewer problems than were anticipated with a new car model. It was said this was because engineers and stylists had less time to tinker with and change things because of the short lead times.

Cardboard templates were made from the car's clay model exterior and the design was transferred to an all-wood model. The wood model was made by a subcontractor, and Gale Halderman reports that when they received the finished wood, it had major flaws and was not in many ways exactly the way the car was designed. Due to the short time frame, they fixed what they could on the model, then "just let it go" and hoped for the best. The first exterior metal parts were made from that wood model. The process was preformed at the Body Engineering Building on Oakwood Boulevard in Dearborn.

In April, the first body sheet metal stampings, called Kirksite stampings, became available and were made according to experimental body tool drawings from the wood model. Kirksite is a cheap-to-produce zinc-aluminum alloy used to form metal stamping molds which can be used on a very limited basis before they are worn out. These dies were made by National Lead Company in Chicago and also possibly a few at Dearborn Steel Tubing Company, according to then-body engineering manager Eric Childs.

Two types of dies were made: one type for hand-pressed stampings, and another type for machine stampings called hard dies. Enough sheet metal was stamped out to produce a very few prototypes, possibly built upon modified Falcon platforms acting as surrogates. The number of prototypes built depended on requirements for crash testing, durability and other investigations.

First Mustang specific metal parts were developed here in the Ford Body Engineering Building in Dearborn (2007 photo). *The author*

Normally, about 50 of these prototypes would be assembled for testing, but because of the expedited time frame used for the Mustang, Childs feels that number was reduced for the new model to maybe as few as 25, but that remains speculation on his part. Hal Sperlich recalls there were even fewer than 25 built, and the first one assembled was a hardtop model. It is doubtful the total will ever be known. Procedure then was that all prototype chassis were crushed after use, and Childs feels sure that was the case here. Crushing would have been done about the time the production line started actual production.

There is other information available that indicates the actual first prototype chassis was built as a hardtop for research to test initial body strength. As the prototypes were running durability track tests at the Romeo Test Track, where most of the early testing was done on the cars, some body metal structures began cracking and breaking apart, according to chassis engineer Bob Negstad. As soon as the car started having these major troubles, the "Make-A-Mustang-Out-of-the-Falcon Sprint" small development group of designers started to scatter. Negstad told me personally this problem almost killed the whole Mustang project. Faith had been momentarily lost in the basic design, except by the real believers in the whole project, like Sperlich and Iacocca. Too much money and time had been spent, and this was a problem that was going to be fixed.

Convention tells engineers that convertible chassis must be made stronger and beefier to compensate for body distortion due to the lack of a metal roof for rigid support. It was only the hardtop prototypes that were having the problem, and it was a smart chassis engineer named Bob Stone who suggested the correct approach to solving the problem: they would build the hardtop chassis with the convertible understructure. That meant torque boxes front and rear, side rail reinforcements, upgrades to the rocker boxes, changes to cross members, and modifications to the floor pans. After they adapted the convertible understructure to the hardtop, with added strength of the steel top in the equation, the breaking-apart problem was resolved. In fact, the hardtop was so stiff that later in the '65 production model run, some of those understructure strengthening components were removed for cost savings.

Also in April, Lee Iacocca's requirement for a new disguised show car to become known as the Mustang II would be acted upon. The car was actually built on a metal prototype Mustang hardtop floor pan assembly. After the middle of May, the Mustang II was under construction, with photos showing the Torino nameplate and the Cougar cat on the grille. It would not be finished until September, when it first wore "Falcon II" emblems and finally, the "Mustang II" emblems.

Excitement was growing by the day for those who designed and were responsible for the creation of the new car. Iacocca was convinced he could sell 200,000 cars in the first year, as opposed to the original proposal of 85,000. He knew he would not have enough production capability at Dearborn, so he embarked upon a plan to have upper management approve converting a second assembly plant in San Jose, California, to Mustang production. If management approved his request, he would be completely hung out to dry if his projections fell short, and would probably have to find employment elsewhere.

He sold his plan and the company approved converting the San Jose facility to include production of the new car beginning mid–July of 1964. The original $40 million

Mustang II with Cougar grille emblem and Torino nameplate. *Ford Motor Co.*

investment had now blossomed to $65 million. He then was looking at new projections of 360,000 cars the first year. An important point to recognize here is that rarely in the scheme of production would the company have the confidence to set aside that kind of manufacturing capacity for an unproven new design. Ford became truly committed at this point to a monumental manufacturing program.

In May 1963, the J. Walter Thompson ad agency started production of a film to introduce the new car to Ford dealers, the press, and ultimately the public. At that time, a prototype vehicle was all that was available to be used in the project. The hardtop used was a one-off, hand-built drivable unit and was assembled at Ford Engineering, not at the Pilot Plant. That prototype was trucked in secrecy to Romeo, Michigan, to the Ford Proving Grounds, where photography of the vehicle was completed. The car wore the Torino badging signifying the current name. The film was named *Torino*. Sample advertisements were printed, and headlined "Torino by Ford" and "Brand New Import from Detroit — Torino." Of course, they were never publicly used or shown, due to the subsequent name change to Mustang.

Lee Iacocca told me, "Henry and I rode together in the back seat of one of those prototype cars. Now we're both tall guys, and Henry didn't like the shortness of legroom at all. We both knew it was too small, but we had agreed to live with it. After all, it was for a family of four with two small kids."

The summer months of 1963 were busy planning months. There were still styling issues to be resolved relating to last-minute changes of interior design levels. It was time for most hardware and body-related items to be in final design for specifications to be transformed to prints for manufacturing. Subcontractors had to be chosen through the bidding process to allow contracts to be drafted and put in place. It was time the afterthoughts of a convertible hard plastic top boot and the option of a removable hardtop be forgotten. It had been decided the convertible would have a manually operated top with a power top option, the top boot would be made of soft fabric, and no removable hardtop would be offered.

Accounting bean-counters wanted only the three exterior colors, red, white and blue, to be available to keep manufacturing costs down. Styling insisted on more colors, and more colors were added to the production lineup. Even though the time was short to the introduction date, small styling changes kept creeping into the original design. But with no time left to tinker, the basic original design sailed through right on to the part-procurement phases. The original design and styling were so pure, they required little tinkering.

Henry II returned from Europe that summer with his family and his yacht on what would be their last vacation together as a family. Upon their return, Anne promptly packed her bags and left for New York City, where she took up residence in a hotel suite. She could no longer deal with Henry's infidelity, but had a bigger problem with a formal divorce. It was forbidden by the Catholic Church.

As reported in *The Fords* by Collier and Horowitz, family members were referring to Cristina, the Italian marriage buster, as the "Dago" and "the Pizza Queen." Now that the family wash was in the press, Henry was relieved and carried on more openly with Cristina during the fall and winter of '63, leaving the new Mustang program up to his man Iacocca to watch over. A formal divorce from Anne was granted on February 12, 1964. Family problems concerning his brother Bill's alcoholism were still ongoing as Henry tried to contain and protect his sibling.

As often happens when there are styling changes en route to production, the basics are changed just enough that the original intentions of the stylists and engineers become distorted and the car does not exhibit the final looks as intended. Corrections and adjustments have to be made. In late July, Automotive Assembly Division documents show there were a number of items contemplated for production which were either deleted, changed or modified by the time production Job 1 appeared.

Some were: the requirement was dropped for the gas filler decorative cap to have an anti-theft cable attached due to its attractive design and potential for theft; planned exterior colors Cherry Red Metallic, Silver Mink Metallic, Medium Blue Metallic, Buff, Peacock and Chrome Yellow were dropped; requirement for a black-only dash crash pad was changed to allow more colors; all optional Rally Pacs would be painted matching colors, not only crinkle black; a four-speed manual transmission would be standard with the 289-cubic-inch 2-barrel carbureted engine; there would be a center transmission hump seat pad accessory and a cloth tonneau cover accessory; rear seat carpet would have a vinyl riser directly in front of the rear seat for both model cars; there would be no floor console available if the car was equipped with air conditioning.

Ongoing surveys for the new car had been conducted throughout the year, but it was in the early fall that a special clinic was held to check the public reaction to the finished product. There were fifty-two couples with preteen children selected as subjects by marketing. A requirement for the selection of each couple was that they owned a single standard-sized car. By screening such a group, the company knew the odds were that they were not likely prospects for the purchase of a small sporty hardtop.

They were brought in small groups to the Design Center showroom, where they were shown the new car and their reactions were measured and recorded. Their responses were enthusiastic, but most said the car was impractical for them as a family. Then a key marketing question was posed to each couple individually. When asked what they thought the price of such a car would be, they overestimated by $1,000 or more. When told the price would be $2,380 or even lower, a strange transformation came over the group.

They were allowed to go back and once again look at the car, and suddenly they began to think of reasons why the car would work for their family and be practical after all. Survey analysts immediately realized a marketer's best dream: "If people overestimate the price of a car compared to what it's going to be, you've got a winner!" said planning manager Hal Sperlich. That particular survey not only confirmed what marketing had thought to be true, but set the stage for the advertising theme for first introduction of the car: simply show the car and its price. The surveyed public was happy, the stylists were happy and Lee Iacocca was happy because now he knew without a doubt he had a winner on his hands.

September

The prototype Mustang II designed for a Watkins Glen introduction was completed in September. It was wearing the final design pony emblem mounted in the grille corral and had the original two-seat Mustang pony emblem with vertical bars on the sides.

During the month, major tool fabrication was completed and production tools were used to stamp out body sheet metal. Although there would still be many changes in the tooling, by late September and early October the very first pilot plant cars were under assembly at an Allen Park, Michigan, Ford plant. Many of the chassis parts used at that plant on these units were factory '64 Falcon production line parts grabbed off the Falcon line. Some of those parts traced on the later pilot cars included unibody frame members, coupe hat shelf platform gussets stamped in early July '63, and torque box components from mid–September.

Ford's Allen Park Pilot Plant was the mini-factory/hand assembly location where many Ford car line pilot vehicles were assembled. Pilot vehicles are hand built chassis that are assembled without the use of assembly line procedures and fixtures. They are hand-assembled using the very first generation metal chassis components to check part fit. At

Ford Allen Park Pilot Plant, Allen Park, Michigan (2007). *The author*

the same time, procedures are developed to determine the sequence of assembly to be used by workers on the factory assembly line. This was where the first post-prototype metal Mustangs came together as an assembly.

If all the design, engineering and part manufacturing specifications were accurate, the pieces were hand-assembled easily into a final vehicle assembly. Perfect fit of all parts was never the case with any new model. As apparent problems surfaced with component fit and compatibility, each questioned part was modified and redesigned for tryout on the pilot vehicle. Once everything fit and worked as it should, more pilot vehicles were assembled using the same developmental process until the "perfect" assembly was created.

Engineer Bob Negstad, who visited and worked at Allen Park frequently, related the part change and modification process. A large conference room adjoins the assembly area where engineers and planners meet to discuss problem parts for these vehicles. If, for example, a front shock absorber strut tower under fender cover doesn't have the mounting holes drilled in proper alignment with the inner fender panel to which it is to be attached, the appropriate engineer notes the corrections, redesigns the drawings, then has a new modified part made. When this modified part is made, it is metal stamped with the letter "X," to show it as a modified part, with a number, i.e. 123, to show which design level change has been incorporated on that part.

No matter how well an engineer choreographs his design, there are always unforeseen problems. One such problem surfaced when the new prototypes were being driven at the Dearborn test track. The addition of a relatively heavy V-8 on the chassis with minimum front-end body overhang was causing handling problems at the higher speeds. Negstad related how the development of a special "Handling Package" helped solve the problem on those more spirited models.

Fall was the perfect time to start the PR machine rolling for the soon-to-be-introduced new car. A selection of national magazine writers, *Time* and *Newsweek* writers included, were invited to Dearborn for an exclusive confidential briefing. The cleverly designed presentation thrust was to accentuate the newly defined emerging youth market in such a way that it would be obvious to the writers that there was indeed a market for the new Ford four-seat sporty car. And the writers picked right up on that concept and spread the news in their publications. Correct priming of the new market had been skillfully accomplished by Iacocca and company.

October

Meanwhile, there were other tasks to be tended to in the fall of 1963. The '63 Mustang II prototype car was completed and plans were frantically being made for its debut. Final preparations were made to send the car to Watkins Glen Raceway in New York to arrive by October 4. Once there the car would be positioned for the Iacocca introduction and track demonstration on October 5. On October 6, Ford Public Relations began releasing details and photos of the car to the press. Iacocca anxiously awaited the report on public perception of the new model, although he already knew what the outcome would be. Input at that time as to how the public would react to the prototype was strictly of a marketing nature. Once again, research on the acceptance of such a car was confirmed based on immediate public reaction.

Major production tooling became available in October. The stamping plant and production type body components could be stamped out to assemble pilot plant cars. No date coded components earlier than December have been found on the first pre-production cars. It appears sheet metal stampings from the October/November time frame were used on the earlier pilot cars only, while final stamping adjustments were made for later cars.

November

In November the formal training program for assembly plant supervisors assigned to the Mustang line would be conducted for several weeks. The economy was good and cars were selling well for Ford and the industry, except for Studebaker, which closed its South Bend plant in December.

The country was in a state of euphoria when, without warning, one of the greatest tragedies of the century hit on November 22 at 12:30 P.M. CST. President John F. Kennedy was assassinated in Dallas, Texas. Infamously, the shooting occurred in a Lincoln limousine convertible designed by Dave Ash. That event threw the country into a period of grieving that would resonate for decades. The baby-boomers were becoming christened in a changing world of politics that would thrust many of them into the nation's further involvement in Vietnam.

December

About 60 days from the planned "Start Pre-production Date" of February 7, the Pilot Functional Program would begin. Body sheet metal stampings from final production tooling was used to assemble in-process chassis at Allen Park. Cars built under that program would be used as further advanced pilot plant test vehicles.

From the research done on Mustang hardtop VIN 100002, it has been discovered that some of these semi-completed chassis were transported by truck to the Detroit Assembly Plant (DAP) to be used as the assembly line pre-production startup vehicles. That was exactly the case with 100002.

Parts production was shifting to high gear in December. Parts date-coded in early to mid–December, such as hoods, fenders and door pillars made from U.S. Steel Corporation materials, have been found installed on pre-production cars 100001 and 100002. These two cars also have several parts with date codes from November, such as door window glass. Part inventories were being shipped from the stamping plants and vendors to the River Rouge facility now in advance of pre-production start in February.

It was time to get some advertising film in the can, and in late November it was planned to secretly take some of the pilot plant cars to Arizona for filming new Mustang commercials. Early in December a casting agency from Scottsdale, Arizona, called one of the dormitories at Arizona State University looking for willing college students to do an advertising photo shoot for Ford Motor Company outside Phoenix. A group of Air Force ROTC cadets and associated Angel Flight girls were recruited, and filming in the desert began in mid–December. There were two Mustangs used, a red with red interior coupe and a white with white interior convertible. Scenes were shot with horses running alongside the skillfully driven Mustangs. There were also still shots made with students in winter snow-ski attire posed next to the Mustangs, as can be seen in early commercials. The publicity machine was becoming finely tuned for the introduction date.

December was a roller coaster of emotions for Henry II as his guilt complex concerning his failed marriage kicked into high gear. He was very concerned not only with how it would affect his children, but also how it would affect the company. As Christmas approached, Henry II and Anne flew on the company plane to Hotchkiss School, where they picked up son Edsel and flew back to Detroit near Dearborn. On Christmas, the parents formally revealed to the three children what they had already known: their parents were getting divorced. There was a strained atmosphere around the Ford household that holiday.

Chapter 14

Let's Build the Mustang

January 1, 1964, began a new year of prosperity in the United States. With favorable economic conditions, Congress enacted income tax cuts. Disposable income rose and the mood of the country reflected confidence and optimism. The year would prove kind to the automotive industry. In the first quarter alone, more cars were sold than at any time in the country's history.

At Ford, production parts for the Mustang were being stocked and inventoried. The engine plants were cranking out both six- and eight-cylinder engines for use in the Falcon, the Fairlane, and the new Mustang. The startup of the Mustang would overload the engine plants and, as Mr. Iacocca told me, "That's why so many of the very first cars had six-cylinder engines, because we couldn't get V-8s built fast enough."

The palette of colors was severely limited for the introduction by the "bean counters." Every new color offered would increase the overall cost to manufacture the cars, so initially only Rangoon Red and Wimbledon White were to be produced. However, Caspian Blue was added, along with a very few others. Coincidence or not, ever notice the three colors used with the pony emblem are red, white, and blue?

On January 21, members of the motoring press were invited to preview the new Mustang and were allowed to drive some of the pilot plant cars on the Dearborn Test Track. Many were the original writers who had been briefed the previous fall in Dearborn about the need for a sporty four-seat car for the new market. To address the group, Lee Iacocca was present with his new Ford Division Special Vehicles Manager Frank Zimmerman, an original Fairlane Committee member. Iacocca's opening remarks included, "We think you're in for a driving experience such as you've never had before. The trick is in finding the right combination of roominess and high style — and that's exactly what we think we've accomplished with the Mustang. Come April, we'll know whether or not we're right!"

The pilot cars driven that day have never been identified by ID number. There is no known listing of all of the Allen Park Pilot Plant Mustangs built. Technically, all pilot cars were built by the first week of February, according to plan.

Most pilot cars built would carry a stamped identification number, similar to a VIN, but with an "S" designation for Pilot Plant rather than the usual Dearborn Assembly Plant (DAP) "F" factory code that is part of the vehicle identification number (VIN). There were two photos taken by Ford that display VINs; the first starts with 4S07 (the

Dearborn technical press introduction to the new Ford Mustang. *Ford Motor Co.*

rest of the numbers are unknown), and the second is 5S08F100000. The first 4S07 VIN indicates a 1964 (the 4) Pilot Plant (the S) Mustang hardtop (the 07). The number "4" is a mystery because there were never any 1964 Mustangs produced. It may have belonged to the first pilot plant prototype chassis created before it was decided that all of the production cars would be titled as 1965 models, with "5" the corresponding and correct beginning number.

The second VIN photo indicates that car was a 1965 (5), Pilot Plant (S), convertible (08), with a 260-cubic-inch V-8 (F). The associated 100000 number is a mystery, as Ford began each model's VIN sequence with the number 100001. During an October 2007 research visit to the Ford archives, I discovered that a series of photo negative packets are missing. Those missing packets may have contained the negatives and other photos of these two and other pilot cars. That information is apparently lost. To date, there have been only seven pilot Mustangs accounted for that display the "S" identification, but none of these cars are known to exist.

Typically pilot cars would be destroyed, and the first six listed were reported on Ford records as being destroyed. Those were: 5S08K100002 (note the "K" for high-performance

V-8 engine), a convertible which was modified later by contractor Dearborn Steel Tubing to become their own design Mustang III concept car and which has since disappeared; 5S08F100006, a hardtop with a convertible "08" body designation; 5S08F100008, -100009, -100010, -100014, all four V-8 convertibles; and 5S07U100012, a six-cylinder (U) hardtop assigned to famous six-cylinder engine performance guru Ak Miller (its fate is unknown). There is a report that 5S07[?]100003 existed as a pilot "test" hardtop that was sent in February to the Alan Mann Racing group in England for initial racing tests. It was reportedly retired by Mann after race tests and used as a parts donor for subsequent cars. It cannot be assumed that missing sequential numbers from the identification list were only for Mustangs, as the ascending 100000 numerical portion of the codes used were assigned to all the different lines of cars that were produced at Allen Park at that time. For example, 4S(43)C100003 might have been a Fairlane 500 two-door coupe (body code 43), not a Mustang.

Chassis engineer Bob Negstad and pilot plant manager Ken Reuther told me the pilot plant was very careful not to stamp any VIN onto chassis that could be used for production vehicles on the Dearborn assembly line. This supports why some chassis assembled at Allen Park arrived at the DAP with no VIN assigned, or even with the "S"

January 1964 proposed grille emblem. *Ford Motor Co.*

factory code: because as soon as an identification number was stamped onto the vehicle, a U.S. excise tax would come due, and that, according to a directive "from above," was to be avoided. The "S" coded cars were destroyed at the end of the program so they could never be sold to the public, according to Reuther.

As late as January, there were still three different horse fender emblems under consideration. Various grille emblems were also still being presented.

Design engineers were finalizing plans to relocate the front seat as well as the accelerator pedal to minimize ankle fatigue. Late changes to the design were costly. In the winter of '63-'64, pilot plant Mustangs appeared on country roads being driven for road tests. These test cars generated immediate interest wherever they went. Whether stopped at restaurants or overnight facilities, the cars were surrounded by onlookers wanting to see "Henry's pony car." There was reportedly an offer from a Texan to buy one of the cars for its build price, an estimated $110,000. The Fairlane Committee's marketing research was beginning to pay off.

When any of these admirers were asked what they thought the car cost, none mentioned anything less than $2,500. The average response was, "about $3,500" and some suggested up into the $7,000 range, relating the car to Ferrari — maybe the horse emblem did it. Additionally, thirteen major-city dealers were given live stage shows to build support for the car. Excitement was growing.

The production start date was near. Ford's financial commitment was up to $65 million and Iacocca could afford no mistakes. His career at Ford was at risk. If the new Mustang hadn't produced a quick profit, he might have started his employment with Chrysler earlier. Gene Bordinat succinctly said at the time, "Lee has brass balls. It took a lot of guts. He was putting his whole career on the line."

February

It is unknown today the exact number of prototype and pilot plant Mustangs that were built. By definition, "prototype" means an original or model from which everything else is copied. That would imply there were only three original prototypes made: a hardtop, a convertible and a fastback. In the real world of Ford development, usually something in excess of fifty prototypes might be built. Not so with the Mustang program. Firsthand interviews indicate that, due to the shortness of time, there were fewer than twenty-five built. Normally, all prototype chassis would be scrapped soon after the first production cars came off the line. Eric Childs, Mustang's body engineering manager, told me that was done with all the Mustang prototypes, with none surviving today.

The first hardtop prototype was hand-assembled at the Ford Engineering Lab on Oakwood Boulevard in Dearborn. Information on assembly of the convertible prototype is unknown, but it may have been assembled at the Engineering Lab or at the Allen Park Pilot Plant. From these originals and follow-on improved vehicles, the pilot plant cars were corrected and modified and the build process determined and documented. The process sheets that direct and guide the details of factory assembly were developed by the Pilot Plant, then printed and distributed to assembly line personnel at the factory.

There were a few chassis, according to engineer Bob Negstad, that were saved after non-destructive testing. They were to be used on the assembly line at startup, and were transported by truck to the DAP, where they were stored on the second floor. One of them became the first pre-production hardtop to receive a VIN, 5F07U100002. Its chassis had been assembled at the Pilot Plant and then transported to the Dearborn Assembly Plant for storage on the second floor in the body pool. VIN 5F08F100001, the first pre-production convertible to receive a VIN, may also have been one of those chassis. All chassis in this identified body pool were relocated from storage to the factory. Thus, as assembly on each was completed, collectively they became pre-production vehicles — the name used to describe those Mustangs that were first down the line on or after February 10 and through March 5.

Mustang floor pans that had been modified from the Falcon were a particular problem. Many did not fit correctly in welding jigs or had stamping defects. Negstad told me when I interviewed him they had a particularly large number of problems with the first stamped floor pans at the Pilot Plant. In the process of trying to make them fit, there were many that were hammered into place at bordering joints and at metal bunch points

Weld over transmission tunnel at rear seat area to repair a crack defect caused by the incorrect stamping process. *The author*

WEATHER
Partly cloudy tonight, low 40.
High tomorrow 52.
(Details on Page 3A)

The Detroit News

FINAL EDITION

MONDAY, APRIL 13, 1964 — 91st YEAR, NO. 235 — THE HOME NEWSPAPER — 2 SECTIONS—40 PAGES — TEN CENTS

Ford Puts Its Racy, Low-Slung Mustang on Display

By JACK CRELLIN
Detroit News Industry Writer

NEW YORK, April 13. — Lee A. Iacocca, general manager of the Ford Division of Ford Motor Co., brought his spunky new Mustang out of the company corral today and put it through its paces for 200 representatives of the nation's news media.

Low and racy in appearance, the Mustang bears many of the characteristics of the Thunderbird line. It brings to five the number of models produced by the Ford Division.

The Mustang will be offered in both hardtop and convertible models and will go on sale in dealer showrooms Friday.

Delivery price in Detroit will start at $2,368, exclusive of local taxes.

However, Iacocca pointed out that the Mustang is being offered with the longest list of options and accessories ever available in a new line of cars.

As an example, Iacocca said, options might include a V-8 engine (164 horsepower), which would cost $116 extra, and an automatic transmission at a price of $189.60.

With these extras, he said, a Mustang hardtop delivered in Detroit would total $2,673.60, exclusive of state taxes.

Adding power brakes ($43.20), power steering ($86.30) and a radio ($58.50) would boost the total to $2,861.60.

Iacocca said an all-out sports enthusiast might add to the Mustang a 271 horsepower V-8 at a cost of $437.80, a four-speed manual transmission at $188 and a clock and tachometer for $70.80, which would push the price to $3,064.60, without local taxes.

Standard equipment includes bucket seats, nylon carpeting, floor - mounted shift for both manual and automatic transmission, all-vinyl interior, padded instrument panel and full wheel covers.

"In the Mustang, Ford actually has created three cars in one," Iacocca said. "Starting with the economical, fun-to-drive basic Mustang, the buyer may select options to give him a sports car for street or competition use or a luxury car geared to either economy or performance."

Standard for the Mustang is a 170 - cubic-inch, 101-horsepower six-cylinder engine with a three-speed manual transmission.

Optional engines offered are the Mustang "260" V-8 at 164 horsepower, the Mustang "289" V-8 at 210 horsepower and the Mustang "289" high performance V-8 at 271 horsepower.

(Concluded on Page 2A)

THE MUSTANG HARDTOP—FOALED BY THUNDERBIRD

Detroit News newspaper headline announcing the new Mustang. *From the collections of The Henry Ford Museum, Ford Motor Co.*

at extreme corners. Additional welds had to be made where the metal was stretched too thin and had cracked. The floor pan on VIN 100002, for example, is of very poor quality and was welded to repair a metal crack caused by improper stamping. More changes had to be made to production stampings to correct those problems.

During restoration of chassis 100002, I found many of the assembly welds randomly placed. Obviously they had not been uniformly made in a jig according to any specifications. Once the factory started using production welding fixtures — these were installed on the DAP line only, and not at Allen Park — welds would be "factory" produced according to rigid specs and could be differentiated from random hand welds such as those from Allen Park.

Ford announced to the press on February 7 it would "introduce a new sports-type car, the Mustang." The cat was out of the bag.

Starting the Line

The Dearborn Assembly Plant had a dedicated assembly line producing only 1964 Ford Fairlane cars. That line was identified as the line which would also build the Mustang. The new model would be manufactured by alternating bodies as they moved down the line. But a mass-production assembly line could not be easily stopped to make changes in jigs, fastener bins, and so forth. The entire line had to be stopped to reconfigure.

On Friday, February 7, 1964, the Fairlane assembly line was shut down to be modified, at a cost of $3.5 million, to permit integrated assembly of both Fairlane and Mustang models. New welding fixtures were installed, equipment modifications to the final assembly area were made, an unusually long 1400-foot engine conveyor was installed, and a new convertible trim area was built. Many parts were interchangeable on the two different model line cars, simplifying production and economizing the build process.

When line assembly was started Monday, February 10, the first pre-production Mustang cars were assembled on that line. Those first few chassis partially completed at Allen

Mustang and Fairlane models being produced in 1964 on the combined assembly line at the Dearborn Assembly Plant in Dearborn, Michigan. *Ford Motor Co.*

Park, pulled from the second-floor body pool and turned into the first pre-production units, were assembled on the newly integrated Fairlane-Mustang line. Factory employee Oscar Hovsepian states he was the first to drive a Mustang off the line. Although Hovsepian has no recollection of the color or VIN of that car, he recalls that it was a hardtop model.

Starting the line required all parts be in place before assembly of the first pre-production cars could begin. Yet a number of parts hadn't arrived in time, so some improvising was in order to get the cars moving down the line. Parts were robbed from other lines for use on those early Mustangs. Negstad refers to those scavenged parts as "floor sweepings," since they were parts very similar to those required, but not yet in stock, and were grabbed from Fairlane or Falcon parts bins and installed on the early Mustangs. For instance, the 100001 convertible has a Fairlane turn signal lever installed. A number of other early Mustangs — the ones that are preserved or correctly restored — also have these commandeered parts installed.

Here's how a typical Mustang was built at the Dearborn Assembly Plant. The process started with a dealer order received by the vehicle scheduling department. A vehicle identification number (VIN) was assigned to the order, and the information, including the build specifications for that car, was entered into an addressograph machine that stamped order data codes on a metal "data plate" which would be installed later just below the latch on the rear edge of the driver's side door. Automated information and data processing equipment were rudimentary compared to today, so the order information was likely

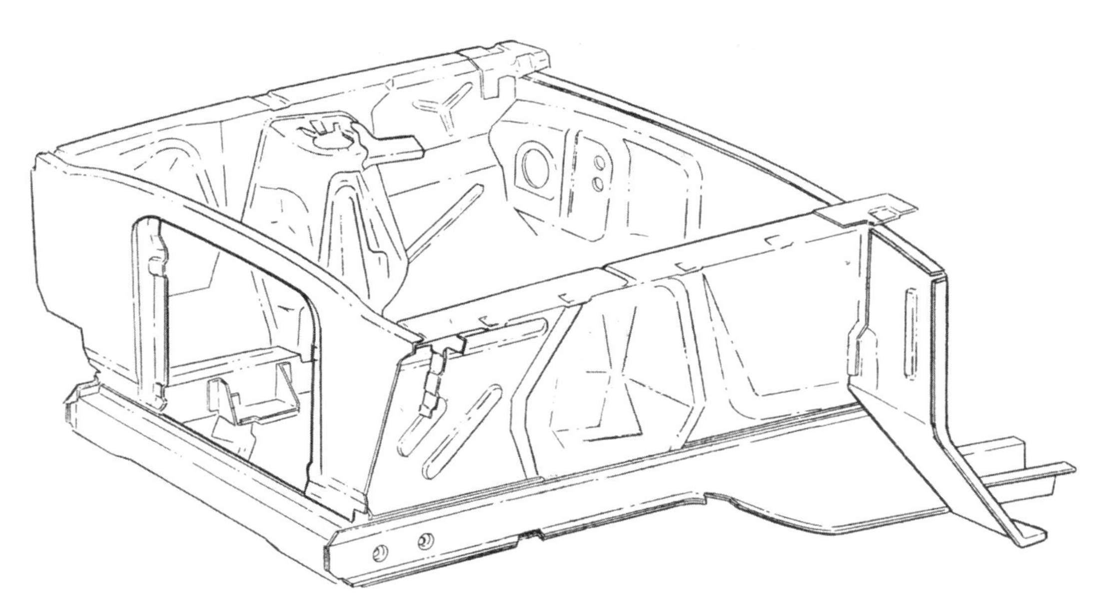

Front Structure Assembly as welded in the "Merry-Go-Round" jig. *Ford Motor Co.*

Assembly as seen in the assembly line welding jig. *Ford Motor Co.*

entered into another piece of equipment so the paper "broadcast sheet" could be printed. The data plate and broadcast sheet were attached together and transmitted to the line setup group responsible for arranging the actual building of the car.

The front structure assembly was built first. It was created by assembling, in the fixture nicknamed the "Merry-Go-Round," the dash panel (also sometimes referred to as cowl or firewall), the left and right inner engine compartment apron panel assemblies, the radiator support, and the outer side frame members. These pieces were spot welded into one single assembly.

One of the first identifications for the new chassis was a four-line set of paint-pencil-applied codes that were written on the right front side of the radiator support near the battery air-vent louvers. The first line of code identified this chassis so workers could correlate the build details specified for that car up until the body-in-white, as it was called, was completed. The actual VIN wasn't stamped into the top of the inner fender apron until just before paint was to be applied to the chassis.

Research indicates the top line numbers on the early production cars were probably assigned in numerical order as the chassis were assembled. Three of the pre-production cars have been located with sequential VINs and coded identification number correlation. However, VIN may or may not have been assigned in numerical sequence to correspond with the top line coded number on these pre-production units. For example, VIN

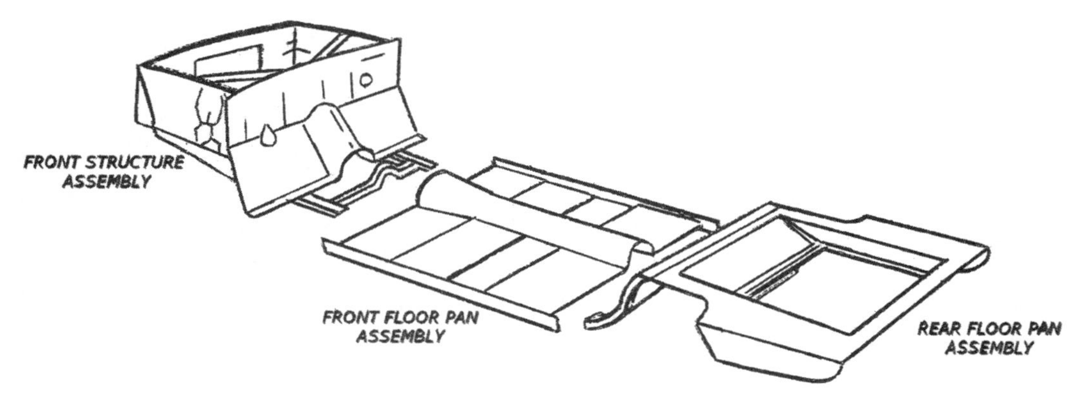

Top **—Four rows of paint pencil build codes etched by age found on radiator support during restoration.** *The author.* ***Bottom*** **—Front and rear floor pan positions for welding to Front Structure Assembly.** *Ford Motor Co.*

5F07U100002 may or may not have been the first chassis down the line, as its first line code would seem to indicate.

The unit was then loaded onto a second buck that was used to align and position the underbody (floor pan or platform) for spot welding to the front structure assembly. Next, the rear floor pan (trunk floor) was loaded into the buck fixture and spot welded to the underbody.

Joining this third component completed the basic platform for the car. On these first cars, Bob Negstad related to me that they had problems with inadequately trained workers not producing secure spot welds. Supervisors examined the moving chassis and, using a screwdriver, tried to break loose a weld that looked suspect. If the weld failed, then the area was rewelded with what was called a security weld to better secure the joint. "Workers didn't like to see their welds fail," said Negstad, because "the worker knew he wasn't doing his job properly and the supervisor was taking notes."

Now a single unit, the chassis was loaded onto a stationary fixture so that the quarter panels, trunk and roof sheet metal stampings could be welded into place. One line worker, Joe Xuereb, recalled installing and spot welding 18 quarter panels per hour at peak rates. The "body-in-white" was now basically complete, and a crane lifted this unit onto a skid (Tool # 49-ZF-3089). The skid was connected to the floor-mounted moving chain and was carried further along the line.

Welding the above floor components to the now one piece Front Structure Assembly with floor pans. *Ford Motor Co.*

Chassis mounted on the assembly line skid as it moves down the line. *Ford Motor Co.*

Each chassis moved next into the paint shop for final preparation and paint. The chassis went into a 100-foot-long paint spray booth where every engine compartment was painted black. Each chassis then moved into another paint section where the Ford-manufactured light gray (pre-production and early cars only) body primer paint was applied, followed by the body color coat. Negstad recalled that after a number of the early cars had been built, it was determined that the gray primer was not allowing proper adhesion with the lighter colors of the new formula acrylic enamel paints that were being used, due to an ultra-violet sunlight interaction. The primer was changed on subsequent models to a rust-colored primer obtained from an outside source.

Following painting, the paper broadcast sheet was taped to the radiator support and the data plate was riveted to the door. The broadcast sheet provided the detail necessary to complete assembly of that car as it moved along the line. In the trim section of the assembly line, interior components — including color-keyed dashboard assemblies, electrical components, glass and other finish items — were installed. No front seats were in place as yet to allow access to the interior for other installations.

Before the front suspension was installed, the left front frame rail was modified by using a torch to crudely cut away a portion of the outer lower lip of the rail. This was done to keep the 14" tire from contacting and rubbing on the frame rail in a hard left turn. Since the rail had originally been designed for the Falcon, which was equipped with 13" tires, this modification was required for the Mustang's larger tires. Engineer Negstad

Top—Chassis painting completed. ***Bottom***—Mustang interior trim installation. *Ford Motor Co.*

related this modification method was used even into the next model year as this fix cost Ford less than reworking the existing die for the frame rail.

The next step required a hoist to lift the chassis off of the skid and let it down to the main floor of the factory (called "body drop"). The chassis was settled onto the rear axle/leaf spring assembly. Once those components were attached to the chassis, the body moved on. At its next stop, it would get its heart — the engine and transmission.

During early Mustang production, some engines, especially the 260-cubic-inch, two-barrel V-8, and transmissions were in short supply. Many of the manual transmissions that were installed were those designated for the Falcon or Fairlane. As an example, 100002 had a Fairlane 3-speed manual transmission and a Falcon Ranchero rear end installed as it was being built. Negstad mentioned that he felt some of the transmissions used were under-engineered for Mustang application, but they were installed nonetheless in order to get cars built. Some of them came out of surplus, salvage and even scrap. They were rebuilt and recertified before being used. Another means of coping with the shortage of transmissions was to ship units in that had been assembled in Mexico.

As a chassis rolled along toward the final assembly area, the front of the car would be completed. Front-end sheet metal, fenders, grille assemblies, and lastly the hood were attached, each in its turn. Wheels, with tires already mounted on them, were placed on their wheel lugs as other items were installed and aligned.

And now, at last, the front seats could be installed, completing the interior. Once

Mustang body being lowered to the main floor for axle installation. *Ford Motor Co.*

Top—**Installing a 260-cubic-inch V-8 engine in the engine compartment.** ***Bottom***—**Hood installation on the completed Mustang front end.** *Ford Motor Co.*

All fluids were serviced prior to the first start up. *Ford Motor Co.*

the car had been serviced with fluids, it was time for its engine to come to life for the first time. If all went well, it rolled off the line as a new and complete Ford Mustang, ready to be shipped to the dealership and taken home by a happy and proud first owner.

Negstad recalled that after all Allen Park pilot cars had been completed, the engineering department had a need for even more chassis for additional pilot plant exercises and testing. Since there were no more Mustangs being assembled at Allen Park, a unit would be pulled as needed off the DAP pre-production line to be used as a pilot car. Thus, it is doubtful that there will ever be a definitive count of the true number of pilot cars as DAP pre-production cars intermixed with Allen Park pilot cars, and vice versa.

Three of those pre-production cars were used, under Bob Negstad's supervision, for installation of independent rear suspensions. They were VINs 5F07F100023, -24 and -31. Although Negstad felt that the independent suspension was superior to the conventional rear-axle design of the production Mustang, the developmental program was never implemented due to the higher cost of components and the end product.

One abnormal pre-production Mustang discovered while researching Ford documents in the Benson Ford Research Library was 5F08F100047. It had had a rumble seat installed. That car is known to exist to this day, but there is no rumble seat. It has been removed, but no one knows why.

Another Mustang with an assembly anomaly is VIN 100148, equipped with a 289-cubic-inch, high-performance V-8 engine. It stood out in Dick Cottrell's memory. He

Completed 1965 Mustangs as they rolled off the assembly line. *Ford Motor Co.*

was a line supervisor and recalled that the exhaust system wasn't a good fit. When the car was started, the vibration of the mufflers against the frame practically "rattled the windows out of the car." It was pulled aside and the exhaust was reworked. This Mustang, even more importantly than all of the cars for customers, had to be right because this one was predesignated for Henry Ford II.

Mustang VIN on the DAP-built cars were exclusive to each model. Even though Mustang was assembled interspersed and alternating on the same line with Fairlanes, each model had number identifications exclusive to it. Odd as it may seem to those unfamiliar with the ways of Detroit, 1964 Fairlanes were being assembled right alongside Mustangs that were all 1965 models. The "4" was the first number in Fairlane VIN sequences and "5" was the first number on Mustangs, and no Mustang was ever built using a 4 as the first character of its VIN. Numbers were not assigned in order to cars coming off the assembly line regardless of model, but rather were in numerical sequence related to each specific model. Fairlane was already about halfway into its 100000 series numbering at the time the assembly of sibling Mustang commenced with VIN 100001.

February 9, 1964, was a Sunday. Many Ford assembly plant workers were, like most Americans, home watching *The Ed Sullivan Show* and the first appearance on U.S. television of a new British rock group. The Beatles made history that night. Some of those Ford workers would make their own history the next day as they began producing Ford's newest model, the Mustang.

The retooled Dearborn plant's assembly line began moving Monday morning, February 10, 1964. Mustang models were interspersed on a very limited basis with Fairlane cars. The actual build dates for those first pre-production Mustangs is unknown because all of them had had a preassigned March 5 (05C) build-date code already stamped on their data plates.

Some of those very first pre-production cars had been designated for a special use. The first formal mention of Mustangs being used as part of the Magic Skyway ride at the 1964 World's Fair had been December 23, 1963. Two months later, Ford documents dated February 6 indicate that twelve of the first pre-production Mustangs had been committed. Another document from this date still referred to them as T-5 models, rather than Mustang by name. The early cars were needed immediately in order for Carron & Co., the business selected to modify and prepare the cars for use on the Magic Skyway, to meet its modification and installation schedule.

The VIN of the first of the twelve Mustangs designated for the World's Fair Skyway ride was 100003. All were convertibles equipped with 260-cubic-inch V-8s and automatic transmissions. VINs 100003, -4 and -5 were Raven Black, -6, -7 and -8 were painted Wimbledon White, -9, -10 and -11 Guardsman Blue, and -12, -13 and -14 Rangoon Red. The first of these sent to New York after modification was 100009, and it was shipped on March 19.

By February 28, 150 pre-production Mustangs had been assembled. Some would be used for display at auto shows in the United States and abroad. Some of them would eventually be sold at retail after use by Ford Motor Company. Initially, all were stored deep within the DAP in a secure, no-access area to assure there would be no unauthorized pictures or leaks.

Late in 2007, I interviewed Don Krencicki, who had been one of the program coordinators. He told me that he had been able to gain access to that secure area of DAP where the group of pre-production Mustangs were stored. He recalled there were around 100 cars, both convertible and hardtop models, and in many colors. Perhaps of greatest significance, he vividly and specifically remembers there were several fastback models in the mix. It has been previously thought that no fastback Mustangs were built until July, when true fastback production began. But, according to Krencicki's recollection, there were a few fastbacks there. Allen Park Pilot Plant prototype manager Ken Reuther can't confirm there were fastbacks there, but he stated to me, "These had to be prototype cars that were run on the DAP line, possibly with tacked-on fiberglass roofs for internal use and evaluation. It was not uncommon to run a prototype car down the DAP line, as this was the fastest and easiest way to get the car painted, etc. Their unique interiors would have been installed afterward by the prototype lab."

In 2006, I learned of one fastback that is rumored to exist in the southeastern U.S. that has a factory-installed hardtop data plate. This may be one of those pre-production fastbacks. Neither Hal Sperlich nor Gale Halderman specifically remembers any pre-production fastbacks being built during the February/March time frame, but each also told me that it was a distinct possibility they existed as prototype cars.

Today, most of those cars used as show cars are identifiable, for each has a build specification known as "Show Car Treatment." As those cars were at the "body-in-white" build stage, the metal seams of the doorjamb and the perimeter seams of the trunk opening were leaded smooth so they would have an improved "no-seam" appearance after painting.

Ford had committed to providing Mustang as the pace car for the upcoming 1964 Indianapolis 500 race. In addition to the pace cars, there was a requirement in the race track's agreement that 100 additional Mustangs be provided as courtesy cars for the festival activities and for the use of dignitaries. Some of the pre-production cars that had been damaged in subsequent testing were reportedly repaired to help fulfill the Indianapolis 500 needs. Holman Moody race-prepared the pace cars and may have been involved in repairing some of the additional Festival cars too.

Pre-production cars give the assembly line workers a chance to learn the build process. Initially only a few Mustangs were interspersed on the line with many more Fairlanes, so there was less risk of slowing the overall production rate of the line during the learning process. It can take several days, working a new model along the line with numerous changes and "tryouts" taking place, before a vehicle representative of desired production quality is achieved. Dick Cottrell was one of the line supervisors then. He told me, "About one of every four pre-production Mustangs was a convertible." Extrapolation of that information would indicate that about 45 of the approximately 180 pre-production Mustangs were convertibles.

The current owner of convertible 100047 forwarded me a copy of an assembly line inspection sheet found in the rear quarter panel of his pre-production car. The form was entirely handwritten, befitting the very early stages of production before the creation of standardized forms for such inspections. This car also still has its prototype Bendix radio faceplate.

Last-minute changes were common. Even the hood sheet metal on the first cars was being modified as cars were being built. The newer design featured a drop flange at the front edge corners. That style hood began to show up on cars built in the first part of April. Ironically, this lip-flange design reverted back to the original no lip-flange style in August of 1964 to accommodate a grille design modification.

March

March 5 was the 64th day of the 1,964th year A.D. At the Dearborn Assembly Plant the transition from pre-production to making full-on production Mustangs was going at lightning speed.

The remaining 30 to 50 pre-production cars were completed by Thursday, March 5, a production-planning date, and all pre-production cars were assigned a March 5 data-plate build date code, regardless of which day or month they had actually been constructed. The last *known* Mustang produced with a March 5 build date was VIN 100178, a Wimbledon White convertible designated for export to England.

All of the pre-production cars, as far as this writer has been able to determine, carry data plate DSO (District Sales Order) codes that are between 81 and 99, indicating sales/usage by Ford of Canada, government, Ford home office (in-house use), American Red Cross, Ford transportation services, or export sales. None of those cars were originally built to specific dealer orders. This group of pre-production Mustangs is truly rare, with many proving to be one-of-a-kind custom-built examples.

Making the assembly line ready for full production of Mustangs and Fairlanes was completed over the March 7-8 weekend. Parts bins were relocated, inventory bins were loaded with Mustang parts and work stations reset. One source indicated that some issues were still being resolved the day before Job 1 rolled off the line. Specifically noted was a resonance problem with the exhaust system and the 260-cubic-inch V-8 installation. An engineer discovered that exhaust brackets for Ford 390-cubic-inch engines, when mounted in a new location on the Mustang chassis, resolved the resonance problems. For the first few days of production, bracket parts from one Ford line were hand-carried to the Mustang line and line personnel were verbally instructed where to mount them.

On March 9, 1964, exactly 571 days after the day in the styling center courtyard when Joe Oros's design had been selected, the first retail production Mustang was a reality. Assembled on that Job 1 day was VIN 100211. It is the earliest known Job 1 day car to exist at the time of this book's publication. It filled an order from Hull-Dobbs Ford in Winston-Salem, North Carolina, a dealer within DSO area code 22. There may have been Mustangs with lower VINs built on March 9, but this is the lowest known to date. No record, written or photographic, has been discovered that indicates the VIN number of the first Mustang to roll off the line on that Monday. The identity by VIN of the first true production/retail Mustang produced will most likely never be known, for one reason. Typically, any internal or publicity photos made by Ford of the first car to roll off the line that first day would picture a pre-production car purposely placed at the head of the line.

For publicity, the public relations department required the Job 1 car be a pre-selected model with the correct photogenic color and top configuration. Exact clones were built of the models selected for first roll-off to reduce the risk the exact car pre-selected as the roll-off car would become damaged before the publicity photos were taken at the ceremony. In the event of damage, a direct substitute would be immediately available as a backup. (This procedure was related by Gale Halderman.)

Data confirms there were approximately 180–200 (plus or minus a few) pre-production Mustangs built at the DAP prior to March 9. And to date, forty-one of these pre-production cars have been accounted for. The specifics regarding some of them are interesting.

One of those was the previously mentioned 5F07K100148 that had been specifically designated for Henry Ford II. Gale Halderman was at the plant on March 9 to observe Job 1 roll-off, and he recalls he had been especially watching for this Raven Black hardtop with a high-performance engine. He had been told it would be the fourth Mustang to come down the line, in keeping with Ford's practice of pre-positioning especially photogenic cars that would guarantee good publicity photos.

Another pre-production car was a red convertible that was delivered on the night of Tuesday, March 10, to the Grosse Point, Michigan, residence of Henry Ford's only sister, Josephine. Her husband was named Walter Buhl Ford II, a coincidental name, since he was not related by blood to any in the "Ford-of-the-blue-oval" family.

It was right and fitting that Josephine, as a principal stockholder of the Ford Motor Company, have one of the latest creations of her company. But with the Mustang's formal, public introduction not scheduled until the following month, the occurrence of the next morning will always leave people wondering.

Wednesday morning, Josephine and Walter's eldest son Walter Buhl Ford III, 20 years old and scheduled to be married the following weekend, strolled into their estate garage and gazed upon the beautiful new Mustang. Without a second thought, Walter III got in, started the car, and drove it into downtown Detroit for lunch at a trendy restaurant. He parked it in a busy lot on Washington Boulevard near the Sheraton-Cadillac Hotel. Be it fate or preplanned, a reporter for the *Detroit Free Press* spotted the car on his way to work. Photos were taken and the new Mustang was revealed to the public on the pages of the *Free Press.* Within a week, those same photos appeared in national weekly publications such as *Newsweek* and *Time.* The whole world (including your author) saw for the first time this new four-seat sporty car from Ford that they called the Mustang.

General opinion and speculation had it that the photo incident was carefully planned to pique the public's curiosity and was not accidental at all. However, when I mentioned this part of Mustang lore to Lee Iacocca in September of 2006, he responded with, "Yeah, that dirty kid Walter Buhl Ford just let our secret out of the bag. It wasn't planned at all. We just had to live with it. But in the long run it never seemed to hurt."

VIN 100465 has a documented build date of March 9. That helps us to know that at least 153 Mustangs (perhaps more) were assembled on that first day of production. Included were 100240, -241 and -242, which were designated to become *the* white Indianapolis 500 pace car convertibles. By that "lucky" Friday, March 13, VIN 100788 had been assembled, indicating that at least 576 cars were built throughout that first 5-day week. By the end of that first week of production, 767 pre-production and production Mustangs had been assembled since the first one that had been built in early February.

One of the very early cars had been earmarked for Mustang Project Manager Don Frey. It was taken to the styling department, where it was reportedly modified to the tune of $100,000. This included a custom red paint job, but not before all of the seams were leaded smooth. The interior was redone in full leather and much of the brightwork was chrome-plated. The car was delivered to his office. After experiencing his first drive, he is quoted as saying, "It's not a wonder it's selling so well. It is an amazing car and looks absolutely beautiful." His passenger, who was none other than Bob Negstad, retorted, "It should have, this car cost Ford a fortune!"

Twelve-hour work days were common on the assembly line, yet Mustangs were being assembled at a slower rate than they were being sold. At full capacity, the line was turning out 75 completed Mustangs per hour, about 1200 cars per day.

The original Mustang (Cougar) Project, as approved, was predicated on building the car integrated with the Fairlane line at the Dearborn Assembly Plant. However, it became obvious early on, based upon actual sales and updated marketing research that projected a demand for over 200,000 cars in the first year, if production of Mustang was going to meet that demand, greater assembly line capability was needed. In mid–June Fairlane production was moved to another assembly plant, and the DAP was dedicated to building only Mustangs. So popular would this pony car prove to be that it wouldn't be long before even the DAP, operating at full capacity, wouldn't be able to produce enough cars to meet demand.

Finding the Holy Grail

Collectors seem to require identification of the "Holy Grail" of their particular collecting world. Usually that "Holy Grail" is defined as "the First One." But sometimes, as with Mustang, this is not such a simple task. For instance, which "first"?

- The first Mustang off the assembly line on March 9, 1964?
- The first pre-production car assembled on February 10?
- The first car assembled as a pilot vehicle at the Allen Park Pilot Plant in the fall of 1963?
- The first drivable prototype car assembled for initial testing in the spring of 1963?
- The 1963 Mustang II prototype?

For certain, each phase had its own first car.

Ford Motor Company has no records that have been found to date indicating which Mustang was the very first to roll off the Dearborn plant assembly line on February 10 or March 9. It is lost in the mist of time which was the "first one built" in any category. What intrigues the hobbyist decades later simply wasn't an important item then. Ford was busy concentrating on manufacturing a new car for the marketplace.

Does it matter if there are no Holy Grail Mustangs? It certainly doesn't lessen the enthusiasm for today's hobbyists. And it doesn't alter the genuine impact this car had, and continues to have, on the automotive world. Our search to better understand how Mustang was conceived, developed and brought to production has brought clarity to the story and even set right some misinterpretations.

The first prototype built in the spring of 1963 was reported to have been a hardtop. The one-of-a-kind 1963 Mustang II was built from one of those prototype hardtop chassis. The chassis used was acquired for that purpose later in the prototyping build stage, so it most likely was not the "first" prototype hardtop chassis assembled. Whether or not it was or wasn't the first one built remains unproved. Presumably all of those other prototype cars were destroyed, and there is no solid information available today regarding any of them.

Which was the first Allen Park pilot car built? No records at Ford give clear evidence of the first pilot car assembled, nor is there a definitive record of precisely how many were

First pre-production 1965 Mustang convertible to be assigned a VIN (Vehicle Identification Number), 5F08F100001. Originally purchased by Canadian Eastern Provincial Airways Captain Stanley Tucker. *Ford Motor Co.*

constructed. As previously mentioned, there is one photograph showing a pilot plant VIN stamped on a fender apron of a hardtop that begins with 4S07. But no chassis carrying a "4" designator were constructed, and that 4-bearing pilot car was reported crushed with the others. There is a photograph of VIN 5S08F100000, which conceivably could have been the first pilot plant chassis. But then again, it is believed all those numbers started with the first car numbered sequence 100001. So what was that car? As there is virtually no doubt it was sent to the crusher over four decades ago, we'll likely learn no more about this pilot car.

1963 Mustang II prototype and the first pre-production 1965 Mustang hardtop with their creator, Lee Iacocca. This was the only time these two famous Mustangs were ever photographed together or together with Lee Iacocca. Photograph taken at the Mustang 40th Anniversary display at the Petersen Automotive Museum, Los Angeles, California, 2004. *The author*

What about those pre-production cars? There had to have been a beginning, a first of the approximately 180 pre-production cars. However, Ford was (and still is) in the habit of not building cars sequentially by VIN. Build sequencing was (and is) dictated by production requirements and build orders as they are passed down from scheduling to the assembly line. Thus 100005 may have been built first. Likewise it is not known whether 100002, the first pre-production hardtop assigned a VIN, was built before or after the first pre-production convertible assigned VIN 100001. For that matter, the identity by VIN of any other of the cars assembled on the first day of pre-production, much less the identity of Job 1, is unknown.

Then too, recall the recollections of Gale Halderman regarding Ford's practice of staging the start of production for photo purposes. Halderman recalled the first car off the assembly line on March 9 was a hardtop model, but he doesn't remember any details of the car. Ford archive personnel have never found any pictures that document Job 1 that day. Halderman vividly still remembers waiting for the Raven Black Mustang that was designated for Henry Ford II, which substantiates that it was one of the 180 or so pre-production cars that were assembled before March 5 and placed for photography purposes on the production line on March 9.

In light of the foregoing, the Henry Ford Museum curator Bob Casey opted to revise the sign and information relating to the Wimbledon White Mustang convertible on museum display bearing VIN 5F08F100001. The sign explaining the car to museum visitors had described it as "the first Mustang built," originally belonging to Stanley Tucker, a captain for Canadian Eastern Provincial Airlines. However, in February 1999, the Henry Ford Museum retitled the exhibit and all references to Captain Tucker's car as "Mustang serial number 1, the first Mustang to receive a VIN number."

Chapter 15

Coming April 17, the Unexpected

Presenting the unexpected ... new Ford Mustang! $2,368 f.o.b. Detroit

Mustang production began on March 9. Formal introduction was scheduled for April 17, 1964. Ford planning called for assembly of at least 8,160 Mustangs before the introduction date so that every Ford dealer in the U.S. could have at least one Mustang in its showroom on opening day. We know from actual verified VIN sightings at least 14,799 Mustangs were actually produced by April 17. The challenge was in implementing the logistics to get timely distribution of the vehicles to the dealer network.

It was time to activate marketing plans. In a brilliant PR move, several weeks before the car introduction, 200 top radio disc jockeys were invited to Dearborn to test drive the new car. Those who took the drive were given car fact sheets and were asked to ad lib on-air about the car. Spreading the word and creating excitement: that's what it was all about, and that's exactly what they did. They may have been influenced by the new Mustang car that was loaned to each of them by their local Ford dealer for a week.

The car market target was in large part college graduates. What better idea than to give college newspaper editors across the country the use of a new Mustang just a few weeks before introduction? A plan was made and it was done. Like the wild horse that it was, Ford marketing was searching for spectacular and unconventional ways to introduce the car to the public.

The Fairlane Committee had envisioned as early as 1961 use of the New York World's Fair as a launch venue for the new Mustang. Don Frey said Iacocca was unbending from the days of the committee meetings on his plan for a World's Fair introduction and would accept no delays. It was an extension of that plan to formalize the car's press introduction at the fair. On April 13, 124 invited media representatives from the United States, Canada and Puerto Rico assembled at the fair. Lee Iacocca himself addressed the group and welcomed them to festivities that would be part of an international introduction of the Mustang. In Europe, 2000 press, radio and TV newsmen were also being introduced to the car.

Following the actual media presentation of the car in New York and after a short marketing film, the Fair group relocated to the nearby Westchester Country Club for lunch, and all were then invited to drive a new Mustang on a 750-mile road rally to Detroit. Each of the 70 cars that were driven on the junket preformed reliably and the durability of the cars was proved. Spectators reported to family and friends after seeing the cars on the road that they loved what they saw. This was just another feather in its cap for the Ford public relations program.

Ford Motor Company had big plans for attendees with a massive fair exhibit that had been in development for several years. They constructed a large rotunda-style exhibition building to be called the Ford Pavilion.

The opening ribbon to the Ford Fair exhibit was cut when a new Mustang rolled through carrying Henry Ford II, Lee Iacocca and Walt Disney. Photo opportunities were plentiful for the Ford PR machine that opening day. Another photo showed a Mustang drive by with Frey, Iacocca and Henry II passing in front of a group of other Mustangs strategically located at the fair.

The first level of the rotunda building would feature all the new 1964 Ford Motor Company cars from each car division, along with some futuristic designs. A ride on the escalator to the second floor would take the spectator to an entry area where he could ride on the Magic Skyway designed jointly by Ford and the Walt Disney Company.

This was an elevated, motorized roadway that would encircle the exterior of the building. Ford company convertible cars were visible from the outside of the pavilion as they moved along the track. Different models from the various Ford divisions were used. As these convertibles traveled along the ride, a built-in audio device would play a description of the car's attributes to the listener in any one of four different selectable languages.

Night lighted 1964 World's Fair Ford Pavilion in New York City. *Bill Cotter, © worldsfair photos.com*

1964 Ford Galaxie convertible enters the Magic Skyway ride with Henry Ford II in the driver's seat and Walt Disney in the rear; the front seat passenger is unidentified. *Ford Motor Co.*

A moving track for full-size automobiles had never been designed before. Henry II asked his good friend Walt Disney to provide the technology for the design and construction of the ride. In the summer of 1963 Disney assigned his project engineer Bob Gurr, who coincidentally had worked for Ford a few years earlier as a stylist, to develop the ride for Ford.

In order to design the fixed platen (flat surface) to be installed under each car, and the track fixed-drive wheels, Gurr had to have all of the size and weight specifications for each type car to be installed on the ride. He went to the Special Process Lab located at the Dearborn Test Track across from the Dearborn Inn in October 1963 to obtain the

Magic Skyway ride can be seen encircling the exterior of the Rotunda. *Bill Cotter, © worlds fairphotos.com*

information. Gurr was taken into a secret area where three developmental cars designated to him as "R" cars were located. Gurr was told these were hand-built cars (pilot cars). He was not allowed to take any pictures and was told not to discuss these cars with anyone. He took the measurements and specifications he needed and left, never to see any similar cars until the evening of April 16. That night, the all-new pre-production special order Mustangs built and modified for the fair were loaded in secrecy on the Magic Skyway ride for introduction to the public the next morning. Among all those Ford company division cars installed on the ride, twelve were the previously unseen Mustang convertibles.

These Mustangs were identically equipped with 260-cubic-inch two-barrel carbureted V-8s with Cruise-O-Matic transmissions and single-key lock sets (same ignition and trunk key). All cars were identically keyed, meaning one key fit all. Contrary to later popular belief, according to Mr. Gurr in my interview with him in 2005, none of the Mustang engines, transmissions or fuel tanks were ever removed for the ride's weight considerations, as was required on other, heavier Ford models, like full-sized Fords, Mercurys and Lincolns. Mr. Gurr emphatically states the only reason for removing these heavy components from the bigger cars was that the ride's push motors would only handle cars up to a certain maximum weight, and the Mustang was light enough not to require the removal of any components.

Lee Iacocca, Henry II and Gene Bordinat chat over a new convertible. Notice the Goodyear branded rear tire. *Ford Motor Co.*

Dr. Martin Luther King and his family were photographed riding in a Mustang convertible on the Magic Skyway later that day, as were other noted celebrities.

There was one more bit of advertising glitz added to the fair's agenda, involving an award that was coveted most by Henry II. It was the Tiffany Gold Medal Award given "for Excellence in American Design." It would be used heavily in advertising. Actually, the award was created just for the Mustang at the request of Ford Motor Company. The Mustang was not selected by Tiffany. Someone in Ford marketing, according to Hal Sperlich in product planning, thought the car was so beautiful it should get an award. Ford marketing manager Chase Morsey Jr. was sent to the famed diamond and jewelry retailer Tiffany's in New York to see if they could persuade the Tiffany CEO Walter Hoving into creating an award for the car. The award would be used in the marketing campaign. Hoving agreed, and after the "appropriate fees" were negotiated, the Tiffany Gold Medal Award was created and presented to Henry II at the introduction ceremonies at the World's Fair on April 13. The Mustang was hands down the winner of the prestigious Tiffany Award, not that there were any other candidates allowed.

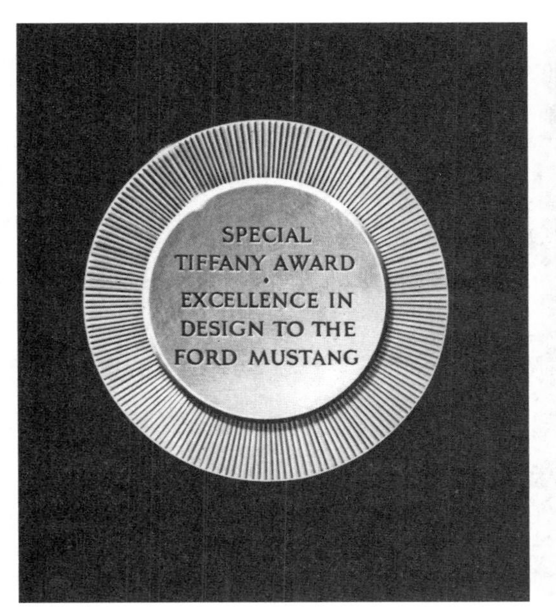

***Left*—Prototype Tiffany Medal.** *Hartman Center, Rare Book, Manuscript and Special Collections Library, Duke University.* ***Right*—Tiffany Gold Medal for "Excellence in American Design."** *Courtesy Peter Grant*

Ford marketing was on a roll, with the public primed for the introduction date just a few days away. Communications technology would take a giant step ahead at the fair with Bell Laboratories' introduction of the touch-tone pushbutton phone, which would replace the rotary dial phone for the world. On an opening date that would coincide with the first public introduction to Kellogg's Pop Tarts and General Mills' Lucky Charms cereal, how could they miss?

Thursday night, April 16, was one for the television viewing history books. The Ford Marketing Department had purchased all of the prime-time programming between 9:30 and 10:00 P.M. on all three major TV networks. This was the first time anyone had purchased time to saturate all the networks simultaneously with their advertising. Ford has conservatively estimated there were 29 million households that viewed Mustang introductory commercials that evening.

Beginning the next day, Friday, some 2600 nationwide major newspapers would carry full-page ads blitzing the public with the new Ford car. Twenty-four national magazines would carry full- and double-page ads spreading the news of the Mustang arrival. *Newsweek*'s opening article said, "Ford is spending $10 million to embed it [Mustang] in the national consciousness like a gumdrop in a four-year-old's cheek." The marketing launch for the new Mustang was the most expensive ever for a Detroit car. The idea to expose the public to the Mustang at every corner peaked to a crescendo on opening day. Mustangs were on display overnight at fifteen major airport terminals, bank lobbies, shopping centers, and two hundred Holiday Inn lobbies across the nation. One might say

GET YOUR FIRST GOOD LOOK AT THE YEAR'S MOST EXCITING NEW CAR... ...THE NEW FORD MUSTANG! TONIGHT—10:00—ON:

NOTE TO PUBLISHER: Strip in correct time per local schedule.

7:00 1
7:30 2
8:00
8:30 3
9:00 4
9:30
10:00 5
10:30 6
11:00
11:30 7
6:00 8
6:30 9
5:00
5:30 0

THE MOST EXCITING THING ON TV TONIGHT WILL BE A COMMERCIAL

GET YOUR FIRST GOOD LOOK AT THE YEAR'S MOST EXCITING NEW CAR... ...THE NEW FORD MUSTANG! TONIGHT—9:30—ON: HAZEL [NBC] · THE JIMMY DEAN SHOW [ABC] · PERRY MASON [CBS]

NOTE TO PUBLISHER: In the event any of the shows indicated are not scheduled locally for Thursday, 4/16/64, please omit mention of same. If only two shows are scheduled center each on one line. If only one show is locally scheduled center in space allotted for program listing.

HAZEL [NBC]
THE JIMMY DEAN SHOW [ABC]

THE JIMMY DEAN SHOW [ABC]
PERRY MASON [CBS]

HAZEL [NBC]
PERRY MASON [CBS]

Magazine ad announcing the carefully choreographed national television introduction of the new Ford Mustang at 10:00 P.M., Thursday, April 16, on the NBC, ABC and CBS networks. *Hartman Center, Rare Book, Manuscript and Special Collections Library, Duke University*

Ford had the undivided attention of the car-buying public with the exclusive mid-year introduction of the Mustang and no competition from other manufacturers. Hal Sperlich said, "This ad campaign was the most successful ever run by any car manufacturer."

Chase Morsey Jr. attributes the success of the initial media ads to their simplicity. A side view of the car was displayed with the "$2,368 F.O.B. Detroit" price inserted above, and words "The Unexpected." Morsey explained to me using the "F.O.B. Detroit" price was an industry first, as an attempt to simplify the real price, discounting transportation costs normally associated with a new car's advertised price in different parts of the country. This allowed Ford to advertise an unbelievably low new car price nationwide.

Car marketing in the mid-sixties kept the element of surprise alive on new models offered by the "Big Three." Great efforts were made to shroud new models in secrecy and out of view from the public until the day of the highly publicized introductory showing. This marketing gimmick created a special array of logistical problems for the manufacturers, shippers and dealers. Marketing wanted at least one new Mustang at every Ford dealer for opening day. The car would be a "low-priced hardtop" equipped with white sidewall tires to coincide with national advertising during introduction. A signed agreement from each dealer required that car to be kept on display (not for sale) until April 25. This required finished units from March 9 through the middle of April had to be transported throughout North America to Ford dealers.

Logistically, it was established that cars destined for Canadian dealers would take the longest shipping time. It is known from the archives of the Ford subsidiary, Ford of Canada, the first two VIN-numbered pre-production cars were sent to east- and west-coast Canadian provinces. These were 5F08F100001, a convertible sent to St. Johns, Newfoundland, and 5F07U100002, a hardtop destined for Vancouver, British Columbia, but misrouted and shipped to Whitehorse, Yukon Territory.

Next priority for the first cars off the pre-production line included the twelve cars designated for use at the World's Fair introduction on the Magic Skyway ride. They were sent immediately to the fair's car modification contractor. Other cars that created logistical problems for shipping were those designated for immediate shipment to Europe and race track testing with Alan Mann Racing in England.

The first pre-production cars were priority units and were handled expeditiously. All of these early Mustangs were carefully cocooned in special covers at the factory so shipments could be kept hidden from public view, including those cars to be forwarded out of the country.

Millions of pieces of advertising were sent out to small-car owners all over the United States. Just days before the Mustang unveiling, big articles on the new car from Ford were carried by national magazines *Life*, *Look*, *Esquire*, *U.S.News & World Report*, *The Wall Street Journal*, and most business and automotive publications. The publicity on the new Ford sporty car just couldn't have reached a higher pinnacle of success; the "Ford boys" were continuing the roll!

Mustangs were unloaded at dealerships at night in fenced-in privacy, so their new styling would not be prematurely seen. I can remember one night, a buddy and I climbed a chain-link fence draped with canvas at a local dealer's lot to catch a glimpse of the new car.

It was a big deal for dads and sons and family members to run down to their new car dealers on introductory day to see all the new models in the showrooms for the first time. It was one of those special things to do on a Saturday morning; race down to the local dealer and see the best Detroit had to offer. It might just have been a time to drool, but of course the dealer's salesmen hoped to turn those impulsive drools into signatures on the bottom line of a purchase contract.

There was something special in the air that mid–April, and families, as if they had been fed a magic potion, were poised to stampede to the Ford dealers on introductory day. On the morning of April 17, every Ford dealer in the country pulled the covers off the new Mustangs in the showroom and near-pandemonium broke out. The first day of retail sales to the public found showrooms packed with spectators waiting to see the new little pony car from Ford. It has been estimated 4 million people visited dealer showrooms that first weekend, an all-time record. But more importantly, they were not only waiting to see the car, but also, even without a test drive, to put down their hard-earned money to buy one.

On that day, orders were placed for 22,000 Mustangs, instantly creating a backlog of 6,900 unbuilt cars. One purchaser slept in his new car at the dealership until the next day when his check cleared and he was allowed to drive it away, and there were many other stories like that one. It was truly hysteria. Don Frey was in the order room at Ford and recalls, "I watched the telex machines register thousands of orders per day during those early introductory days." At one point, it was reported that Ford had a three-month waiting list for their new car. A Detroit-critical *Consumer Reports* magazine article first reported sales by noting "Mustang's almost complete absence of poor fit and sloppy workmanship in a car being built at a hell-for-leather pace."

The first known retail sale of a new Mustang was on April 14, 1964, three full days before the car was formally introduced. The car was VIN 5F08F100001, the white V-8 convertible now recognized formally as the first pre-production convertible assigned a VIN. It was purchased by Canadian Eastern Provincial Airlines Captain Stanley Tucker from the George Parsons Ford dealership in St. Johns, Newfoundland, Canada. That car was not meant to be immediately sold, definitely not in advance of introduction day. When Capt. Tucker saw the car prematurely at the Ford dealership, he made an offer that could not be refused to one over-zealous salesman, and Capt. Tucker was the first on his block to have a new Mustang. The car that was not to be immediately sold got away until the spring of '66, when Ford repurchased it from Capt. Tucker with an offer of a new 1966 Mustang convertible of his choosing. That 100001 car is located today in the Henry Ford Museum in Dearborn, Michigan.

The first pre-production Mustang hardtop assigned a VIN did not see its original retail sale until one year later in the spring of 1965. This car, VIN 5F07U100002, a Caspian Blue six-cylinder hardtop, is owned by this author, and is recognized as the first pre-production hardtop. Shipped to the Whitehorse Motors Ford dealership in Whitehorse, Yukon, Canada, it served there as a dealer demonstrator for nine months. Since the Yukon is quite mountainous, the car was reported difficult to sell with its small 101-horsepower motor. Research proves this car chassis was assembled at the Allen Park Pilot Plant and inserted at the body-drop phase of the pre-production line on February 10, 1964, for final assembly. The car had thirteen unknowing owners between Yukon and Alberta, Canada, and then Wyoming and California before it was located by this author and purchased in 1997.

On April 17, 1964, the first domestic retail Ford Mustang sale to the public was formally recorded. Mr. Iacocca reported to me that sale was consummated by a Boston, Massachusetts, Ford dealer. The VIN is unknown today.

However, recently publicized sale documents associated with the first purchase of a first day of factory production (09C) convertible, VIN 5F08F100212, show on the original Florida dealer sale invoice this car was sold retail and delivered new to its first owner on the evening of April 16, 1964. This is one day before the first Mustangs were to be sold to the public, according to Ford directives. If there are no other documented Mustangs that were sold and delivered domestically on or before that date, it would make this car the "first production Mustang sold and delivered to a retail customer in the U.S."

Lido Anthony Iacocca's own celebrity also was launched on April 17, 1964, and would follow him to this day. This would be a vintage year in several ways for the Iacocca family, as second daughter Lia would be born in July, the same month the second Mustang production facility would be started in San Jose, California.

With the offer of a multitude of options, a marketing coup had been created. Out of the 400,000 Mustangs built in that first year, it was almost possible to order a one-of-a-kind option-oriented Mustang of your own — and that was a major appeal to the buying public. "The Mustang is a practical and economical car, yet, it was designed to be designed by you," claimed advertising brochures and ads. "You can make your Mustang into a luxury or high-performance car by selecting from a large but reasonably priced group of options." The theme carried out in the advertising program was simple: "Whoever you are, whatever you do, this is a car that will make you feel special." And it did.

Lee Iacocca and Don Frey with a specially licensed 1965 Mustang hardtop denoting "417 by 4-17." *Ford Motor Co.*

Within six weeks of its introduction, the Mustang was the top-selling compact-sized car and ranked seventh in sales among all U.S. nameplates, regardless of size. Within four months, more than 100,000 Mustangs were sold, and in its first 12 months on the market, 418,812 cars were sold, a new all-time record for a first-year entry car. The old record was set by the Falcon, and it was Iacocca's goal to beat that record with the Mustang. The Mustang in the photo above displays a license plate "417 by 4-17," indicating more than 417,000 units were built before the one-year anniversary of 4-17, April 17. Actually 1638 more cars were built than the Falcon in the same time period, thereby shattering the record the Falcon had set. This was amazing for a Detroit car that was first projected to sell 80,000 cars in its first year.

Every public relations manager representing a company that has a product to sell knows the unwritten rule that no product likeness appears for free on the cover of a major national news-oriented magazine. Yet through the never-ending efforts of the genius Ford PR managers Walter Murphy and Robert Hefty, that first week after public introduction was so spectacular that the cover of *Newsweek* and *Time* magazines had the new Ford Mustang and Lee Iacocca's name and photo splashed exclusively over the entire cover. "That had never been done before by any magazine, let alone two of them simultaneously," Mr. Iacocca told me in one of our conversations. He credits those cover features alone with an additional 100,000 sales.

Ford Motor Company hit a home run with their new sporty car called Mustang.

The legend had begun.

Chapter 16

ANALYSIS OF THE MUSTANG PROJECT

> The Mustang has written a unique chapter in the history of auto making and marketing.
>
> Ford Corporate Studies

"We were very fortunate that we hit at exactly the right time," said Iacocca.

> In 1964, it was a euphoric time; I mean, they were even cutting taxes. It was just dumb luck that we had the world's biggest showroom — the New York World's Fair — as a launching pad. It was the combination of the World's Fair launch, the fact that we did have a rather unique and different car, a realization that the youth market was bulging, and most of all an economy that was really being heated up by the government's cutting taxes and telling people to go out and spend some money. With those ingredients, it would have been hard not to succeed.

All of the first production Mustangs built from February 10, 1964, through July 31, 1964, were titled as 1965 model year cars. Those cars became known to most hobbyists as 1964½ models, due mainly to their midyear launch in '64. However, even though Ford says there never was a 1964½ model produced and no associated Ford documents or advertising exists referring to a '64½ model, there is evidence to the contrary. A Shop Manual Supplement was printed March 16, 1964, by Ford Motor Company for use by dealer service departments for the "1964" Mustang. It does not refer to the car as a 1964½ or 1965 model, but as a 1964 model. Here is one case where Ford Motor Company referred to the car in one of their printed manuals as a 1964 model car. Is it accurate to say, then, that any Mustang produced prior to March 16 should be referred to by collectors as a 1964 model? You decide.

The '65 Mustang was directed to buyers of four different audiences: (1) two-car families that had some surplus cash to spend on a new car purchase, (2) young drivers with hardly any money to spend but who wanted a new car, (3) women who wanted a car that would be easy to maintain, and (4) the sporty set in search of a new toy. This single car was carefully orchestrated with marketing aimed squarely at these four previously separate market niches.

"With unprecedented success, the original 'pony car' captured the enthusiasm of what became known as the Mustang Generation, car buyers of every age who found in it both economy and style. In fact, it broke previous first year Ford sales records with 418,812 units." This is the way the phenomenon was described in a Ford Corporate Studies

College Resource paper, No. 10 (undated), published by the Educational Affairs Department of Ford Motor Company.

The *Consumer Reports* 50th Anniversary magazine of April 2003 reported the 1965 Mustang was ranked among the nation's 50 top products between 1936 and 1986.

J. Mays, group vice-president of design and chief creative officer at Ford Motor Company, was quoted in a 2005 Mustang road test article (Familycar.com) as saying, "The Mustang is a legend, an icon. It has long since ceased to be an automobile — it is a source of national pride."

Car Life magazine in 1964 said, "A market which has been looking for a car has it now. It is a sports car, a gran turismo car, an economy car, a personal car, a rally car, a sprint car, a race car, a suburban car, and even a luxury car. The car may well be, in fact, better than any domestically mass-produced automobile on the basis of handling and roadability and performance, per dollar invested."

What a summation for all those innovations that went into the Mustang! In the first two years, the Mustang generated net profits for Ford of $1.1 billion (in 1964 dollars). On February 23, 1966, less than two years into production, the 1,000,000th Mustang rolled off the assembly line. Not only did the car sell in record numbers with a base price of $2,368, but customers were spending an average of $1,000 more on options and accessories. Over 80 percent ordered whitewall tires and radios, 71 percent V-8 engines, 50 percent automatic transmissions and every tenth car had a "Rally-Pac" clock and tachometer package. There were engine, transmission, brake system and wheel style and size choices. There were interior color, material and style choices as well as the choice of a floor console, bench seat with arm rest and a simulated wood deluxe steering wheel. And the list went on.

1964 Mustang Shop Manual Supplement. *The author*

There were just a bundle of choices available, and maybe that's what dazzled the buyers. The Mustang had its own chameleon-like personality. There was no such thing as a typical Mustang. An options galore menu made the car very attractive across a broad range of buyers who were willing to spend big in the lucrative option marketplace to build their own car the way they wanted it. A quick survey showed the average buyer was 31 years old and one in every six was 45 to 54 years of age, meaning the car also was having broad appeal outside the targeted group. The market identified by the Fairlane Committee was hit in the bull's-eye.

The ultimate test of success occurs when vocabulary is changed and "trinket" producers flood the market with gizmos related to the product line. Stores began retailing Mustang shirts and pants, sunglasses, boots, earrings, key chains, hats, money clips and model cars. The word "Mustanger" became a new word in our language, defined as a "fun-loving person who drove a Mustang." Mustang clubs began to form across the country like a rapidly spreading fire. These ancillary items and organizations mirrored aftermarket devotion to the car line, a degree of success every marketing man dreams about. The halo effect on other Ford products, both then and now, has been immeasurable with its importance beyond financial success, in a way that reflects customer loyalty to Ford Motor Company products.

If you look at many published pages on world history for the date of April 17, 1964, you will find listed, along with other important events on that date, the introduction of Ford's new Mustang to the public. Historically, the launch of the car has been noted forever in the history books. Try to find that same information on the General Motors Camaro or Firebird, or on Chrysler's Barracuda, or any other but a few modern-day vehicles made in the USA.

At the pinnacle of the automotive design world were several awards meant for design excellence. One of the most prestigious recognitions a designer of an automobile could have bestowed was the Industrial Designers' Institute Award for excellence in design. Each member of the Mustang design team of Joe Oros received that award ceremoniously.

Hollywood immediately took notice of the new Mustang. Some of the early buyers included Frank Sinatra, Debbie Reynolds and George Raft. Others who also joined the cult included Sonny & Cher, Jay Leno, Lindsay Wagner, Jackie Cooper, Reggie Jackson, Debbie Boone and President Bill Clinton.

Contrary to popular belief, Lee Iacocca was never given ownership of a Mustang by Ford Motor Company. He was given the use of several Mustang executive cars from time to time, but ownership was never transferred to him. His wife Mary bought him a used 1965 convertible in 1984, which he still owns.

It would take General Motors 40 years to sell one million Corvettes. It took Ford less than two years to sell one million Mustangs.

Sales were beyond expectations and Fairlane car assembly was separated from the dual use assembly line at the DAP to provide increased Mustang production. The 12-hour work days produced Mustangs at a rate slower than the rate at which they were being sold. This was viewed as a golden precursor, and indeed, orders would be backlogged for months. The assembly line at full capacity was turning out 75 completed Mustangs per hour, and about 1200 cars per day. Still more production capacity was needed. Iacocca quickly geared up factory production to support those kinds of sales by using the San

Jose, California, assembly facility. That production began on July 13, 1964, and soon would be at capacity, when more production would commence at a third Ford facility in Metuchen, New Jersey.

By 1968, there were more than 4,100 different makes of autos that had been produced in America. Obviously, most of them weren't successful. The Mustang was simply the right car at the right time.

The streets and the highways became filled with them. It was a success because it proved to be right for a vast, new untapped market. And part of that market, targeted for the first time by Ford, was the female buyer. Advertising taglines like "Mustang has become the sweetheart of the Supermarket Set" are credited with creating tremendous sales success within this segment of the newly defined youth market. The Mustang set a trend affecting manufacturers for the next 45 years.

In his book *Where Have All the Leaders Gone?*, published in 2007, Iacocca says, "They [meaning baby-boomers] responded to the Mustang because it was a beauty, it had power, and it had an identity. It became a cult car. To this day when I go to the classic Mustang shows, those now-aging baby boomers treat me like a rock star. That car meant something to them." And he was right about that perception. I know. I was there with him.

When it was all said and done, Iacocca, Frey, Sperlich, Halderman and their team had created a car totally in tune with the times and there was indeed a legend created called the Ford Mustang. All known Mustangs built on the assembly line on the first day of pre-production in Dearborn, Michigan, with the exception of three black ones, were painted red, white or blue.

And America became a better place.

Epilogue

Many books have been written about the car named Mustang. This book was about a group of men who set out to manufacture a car that would help further their employment with the mother hen, Ford Motor Company. After reading the book you should have a clear image in your mind about how all these individuals came together in a certain work environment at a magic time in history.

To put these events into perspective, it is important to remember that from the time Henry Ford II took control of the company at the end of World War II until the creation of the first production Mustang prototype was a mere 18 years, the same length of time one has aged by graduation from high school. Major styling and engineering advances took place in a very short time. When that first Mustang was sold, it started a new era in automobile building in America, one that would change our society forever.

Although it would not be possible to list the name of every person who was involved in the original project and further expound their contributions, there were many. Large, small, tall, short, Catholic or Protestant, man or woman, factory worker or blue-collar worker, many a person contributed his or her own personal talent to the success of the Mustang. The faces highlighted in this book were the Goliaths in their own fields. Contrasted to today's politically correct work scene, interestingly, there were no women on the team's first or second tier. Those not highlighted or mentioned beyond those two tiers were, in large part, the worker bees. There can't be a hive if the workers don't all work with synergy to create the end product.

To all those individual personalities and talents, we owe a great deal of thanks. In America, great companies survive because they produce great products. Producing great products guarantees employment to the workers. And all workers understand that without the great products they create, there would be no employment.

What did these people all have in common? They all went to work to bring home the money to pay for the food that sustained their families. The most important thread in our society is family. At the end of the day, we leave our problems at the workplace and go home to the tranquility of the family unit we have created. And the icing on the top of the cake is to know that we have helped to create something good from our productivity. To all those involved in that original Mustang project team, we can say thank you. You gave our society something important that endures, more than just a car.

As a point of interest, from Mr. Iacocca's point of view, he looks back on the creation of the Mustang and says, "Sperlich and I hit the baby boomers once, and then Sperlich and I came back and did it all over again at Chrysler with the creation of the extremely successful Chrysler minivan concept in 1984, just 20 years later, when the boomers were having children of their own and needed larger cars, and they're still building them!"

It has been this author's distinct honor to have become a friend of Mr. Iacocca beginning in 2000. He has graciously over the years made himself available for photo opportunities with my first pre-production Mustang hardtop. He is an honorable, approachable gentleman who at the age of 85 still has a mind as sharp as a tack and is respected and still called upon by world leaders and businessmen for advice. In 2006, it was my honor to volunteer to assist in developing the Lee Iacocca Award, sponsored by the Iacocca Foundation. The award, presented nationally to an individual selected by his or her peers for "dedication to excellence in preserving an American automotive tradition," was developed in the "give-back" spirit Mr. Iacocca fully supports.

Mustang Project Manager Don Frey sometimes is not given sufficient credit for the instrumental role he played in the development of the car. Every executive who conceives a project would select a project manager he could trust and depend on to see that project through to fruition. For a busy executive like Iacocca, who had to devote much of his time to oversight of all of the daily operations of the Ford Division, and not just the Mustang, he placed his trust in the very capable Don Frey. At 39 years old and with a doctorate in engineering, Frey was an innovator and worked well with Iacocca. At the end of the day, it was Frey who oversaw the completion of the Mustang project. He got the job done. Although not thought of by many as the "Father of the Mustang," he certainly was, and still is, at the head of the list of contenders.

Hal Sperlich's supreme intuition, which lent the required insight into an economically viable car, also places him at the top of the list of creators. Had he not presented the use of the Falcon chassis as a basis for the car, the Mustang conceivably would not have been built at that time. As the special project manager, he skillfully guided the entire program through to completion of the first production cars.

Whatever happened to the oh-so-successful design of those original model Mustangs and why was it changed? Stylist Gale Halderman probably summed it up best when he said, "The dilemma that the designers and the company had after the first one was that they didn't know what to do with it. They really didn't want to change it. They didn't want to touch it. But they also knew they had to. It had to be modified — be refreshened. We had heard that GM and Chrysler were both doing cars that might compete against it, so we had to freshen the Mustang."

Well, in this author's mind, that's a subject upon which to dwell. Do you take a chance on tinkering with success and perhaps destroy that success, or do you wait to see if in fact the competition does take away a degree of the sales success? As many car collectors say, "If it ain't broke, don't fix it." In this case a few more good years with the original design might have continued to provide great sales numbers. We'll never know.

We do know the Mustang was created from a unique set of circumstances. Many involved in the initial project viewed the program as a way to relieve Ford Motor Company of its old-school image. It was a dedication to development of the proposal which became a labor of love for something new these men held throughout the program. Conveyed to me through interviews with the workers building those first cars was that they all knew they had a winner on their hands. Mr. Iacocca told me in the spring of 2007, "The years the Mustang project was developed were really good times for me. It was a great group of guys to work with."

As I interviewed the many unique builders for this book, there were several things that became apparent to me. There are many aspects to creating a newly conceived car. Because an individual pours his heart into his piece of the work, there is easily developed a sense of "creation" from his point of view. As some may claim to have been the Father of the Mustang, in fact the creation was the collective result of the labors of love each gave to the project. It is not possible, in this author's opinion, to say there was a definitive "Father of the Mustang." The Mustang was created through the synergy of the entire team. As they have aged, most now in their eighties and nineties, many have remained strong friends and maintain their unique relationships at morning breakfasts, retiree meetings, and their own social get-togethers. They all share in a once-in-history distinctive creation of which each was a part. They sense a satisfaction that what they did at that place and time in the life cycle probably cannot be created again. And every one of them, to this day, talks with feeling when sharing his experiences. They were — and are — the Mustang.

Ford Motor Company continues in operation as this epilogue is written, though not in the best of times for a domestic automotive manufacturer. The Mustang is still a mainstay for Ford with more than 9 million built. The Mustang 50th anniversary will be celebrated in 2014, a tribute to all of these people who gave us something truly unique by any standard, the Ford Mustang.

Bibliography

Printed Sources

Automobile Quarterly 9, no. 1 (Fall 1970).

Brinkley, Douglas. *Wheels for the World.* New York: Viking, 2003.

Car Life magazine, 1964.

Clark, Holly. *Finding My Father.* Rusk, Texas: ClarkLand Productions, 2006.

Collier, Peter, and David Horowitz. *The Fords.* New York: Summit Books, 1987.

Consumer Reports 50th aniversary issue (April 2003).

Detroit News, April 13, 1964, from the collections of Ford Research Center.

Editors of *Consumer Guide. Mustang Encyclopedia.* Lincolnwood, Illinois: Publications Intl., 1982.

Farrell, Jim, and Cheryl Farrell. *Ford Design Department Concept & Show Cars 1932–1961.* Self-published, 1999.

Gunnell, John. *Mustang: The Affordable Sportscar.* Iola, Wisconsin: Krause Publications, 1994.

Henshaw, Peter. *Mustang.* Edison, New Jersey: Chartwell Books, 2004.

Iacocca, Lee. *Iacocca: An Autobiography.* New York: Bantam Books, 1986.

_____. *Talking Straight.* New York: Bantam Books, 1988.

Leffingwell, Randy. *Mustang.* Osceola, Wisconsin: Motorbooks International, 1995.

Lewis, David L. *Ford Country 2.* Sidney, Ohio: Amos Press, 1999.

The Mustang Story: Ford Corporate Studies Series Number 10. Educational Affairs Dept., Ford Motor Company.

Smart, Jim, and Jim Haskell. *Mustang Production Guide.* Vol. 1. N.p., In Search of Mustangs, 1994.

Witzenburg, Gary L. *Mustang!* Princeton, New Jersey: Princeton Publishing, 1979.

Wright, Nicky. *Mustang Anniversary Edition 1964–1994.* London: Multibooks, 1994.

Other Sources

Benson Ford Research Center, Dearborn, Michigan.

Detroit Historical Museum, Detroit, Michigan.

Ford Archive Services, Ford Motor Company.

Ford Communication Services, Ford Motor Company.

The Henry Ford and Greenfield Village, Ford Motor Company.

The National Automotive History Collection of the Detroit Public Library, Detroit, Michigan.

Owls Head Transportation Museum, Owls Head, Maine.

University of Michigan at Dearborn, Science and Technology Studies Program, Dearborn, Michigan.

Interviews

Ash, Dave (deceased). Ford Styling Department. February 28, 2008. Telephone with Rosemary Ash, wife.

Buckley, Norbert. Dearborn Steel Tube Co. manager. October 3, 2007. Telephone.

Casey, Bob. Curator, The Henry Ford Museum. February 3, 1999. In person.

Childs, Eric. Ford Body Engineering manager. October 11, 2007. Telephone.

Cottrell, Dick. Dearborn Assembly Plant supervisor. July 3, 2006, Telephone.

Doss, Larry. Dearborn Assembly Plant worker. June 15, 2003. In person.

Eby, Jack. Dearborn Assembly Plant manager. October 4, 2007. Telephone.

Frey, Donald. Ford Division president. May 18, 2001. By letter.

Gurr, Robert. The Walt Disney Co. March 7, 2005. In person.

Halderman, Gale. Ford Design Studio. October 3, 15, 18, 2007. In person. October 20, 2007, and June 6, 2009. Telephone.

Hutchins, Paul. Ford Mustang Interior Studio. June 2, 2009. Telephone.

Iacocca, Lee. Ford Division president. Various dates, 2001–2007. In person.

Krencicki, Don. Ford program coordinator. June 15, 2003. In person. October 3, 2007. Telephone.

Lunn, Royston. Mustang I project manager. March 27 and April 14, 2008. Telephone and by correspondence.

Mason, James. Dearborn Steel Tubing Co. April 17, 2002. Telephone.

Morsey, Chase, Jr. Ford Division marketing manager. April 28, 2008. Telephone.

Moskowski, Lee. Lincoln-Mercury division manager. April 28, 2008. Telephone.

Najjar, John. Ford Pre-production Vehicles Studio. March 3, 2008. Telephone.

Negstad, Robert. Ford Suspension design engineer. September 5, 1998. Telephone. April 25, 1999. In person.

Passino, Jacque. Ford racing manager. March 24, 2008. Telephone.

Quick, Eldon A. Dearborn Assembly Plant worker. June 14, 2003. In person.

Reuther, Ken. Allen Park Pilot Plant prototype manager. June 7, 2009. Telephone.

Richards, Jess. Ford Metal Stamping Division. October 14, 2007. Telephone.

Ruth, Jim. Ford Styling Department. June 15, 2003. In person. November 5, 2007. Telephone.

Schumaker, George. Ford Styling Department. June 1, 2009. Telephone.

Sperlich, Harold. Mustang program manager. November 6, 2007, and September 2, 2009. Telephone.

Uzeilli, Allesandro. Ford Global Brand Entertainment. August 20, 2008. In person.

Xuereb, Joseph. Dearborn Assembly Plant worker. June 14, 2003. In person and by telephone.

Index

Index

Index